ISBN 978-0-282-69329-9
PIBN 10028253

WOOLWICH — Royal mil.

MATHEMATICAL PAPERS

FOR ADMISSION INTO

THE ROYAL MILITARY ACADEMY

FOR THE YEARS

1894—1903

EDITED BY

E. J. BROOKSMITH, B.A., LL.M.

ST. JOHN'S COLLEGE, CAMBRIDGE; INSTRUCTOR OF MATHEMATICS AT THE ROYAL MILITARY ACADEMY, WOOLWICH

London

MACMILLAN AND CO., LIMITED

NEW YORK: THE MACMILLAN COMPANY

1904

MATHEMATICAL EXAMINATION PAPERS

FOR ADMISSION INTO

Royal Military Academy, Woolwich,

JUNE, 1894.

OBLIGATORY EXAMINATION.

I. EUCLID (BOOKS I.—IV. AND VI.).

Ordinary abbreviations may be employed; but the method of proof must be geometrical. Proofs other than Euclid's must not violate Euclid's sequence of propositions. Great importance will be attached to accuracy.]

1. The opposite sides and angles of a parallelogram are equal to one another.

2. The complements of the parallelograms which are about the diameter of any parallelogram, are equal to one another.

A point E is taken in the side AB of the parallelogram $ABCD$ and ED and EC are joined; prove that, if the line HK parallel to AB cuts ED and EC in F and G respectively, the parallelogram $AHKB$ will be double of either of the triangles EDG or ECF.

3. If a straight line be divided into two equal parts and also into two unequal parts, the rectangle contained by the unequal parts, together with the square on the line between the points of section, is equal to the square on half the line.

4. Divide a given straight line into two parts, so that the rectangle contained by the whole and one of the parts may be equal to the square on the other part.

The line AB is bisected in C and produced to D, so that the square on CD is equal to the sum of the squares upon AB and BC, prove that the rectangle AD, DB is equal to the square on AB.

Express the ratio of BD to AB algebraically.

5. State, *without proof*, the difference between the square on one side of a triangle and the sum of the squares of the two remaining sides.

A point P is taken on the circumference of the circle APB, whose centre is O, and with P as centre, another circle, QAB, is described, cutting the former in A and B. If QM be drawn from Q, any point on QAB outside of APB, perpendicular to AB produced, prove that twice the rectangle of QM and OP is equal to the difference of the squares on OQ and OA.

6. The angle at the centre of a circle is double of the angle at the circumference on the same base, that is, on the same arc.

7. The opposite angles of any quadrilateral figure inscribed in a circle are together equal to two right angles.

8. Describe a circle about a given triangle.

The straight line AB of given length moves so that its extremities are respectively upon the two fixed straight lines OC and OD meeting at O. Prove that the centre of the circle circumscribing the triangle OAB lies upon the circumference of a circle whose centre is O.

9. Inscribe an equilateral and equiangular pentagon in a given circle.

10. If the vertical angle of a triangle be bisected by a straight line which also cuts the base, the segments of the base shall have the same ratio which the other sides of the triangle have to one another, and if the segments of the base have the same ratio which the other sides of the triangle have to one another, the straight line drawn from the vertex to the point of section shall bisect the vertical angle.

11. If four straight lines be proportionals, the rectangle contained by the extremes is equal to the rectangle contained by the means; and if the rectangle contained by the extremes be equal to the rectangle contained by the means, the four straight lines are proportionals.

12. The diameter BCA of the circle $APQB$ whose centre is C, is produced through A to O, and from O the line OPQ is drawn, cutting the circle in P and Q, prove that the triangles OPA and OQB are similar, and prove that if the circle circumscribing the triangle PCQ meets OC in D then

(1) The point D will be fixed for all directions of the line OPQ;

(2) The ratios $OA : AD$ and $OC : CP$ will be equal.

II. ARITHMETIC.

[N.B.—*The working as well as the answers must be shown.*]

1. Find the value of

$$\frac{5\frac{1}{7} - 3\frac{5}{14}}{2\frac{1}{2} + 3\frac{3}{4}} \times 79\frac{1}{2}.$$

2. Divide ·0468 by 29·25.

3. What is the interest on £775 for $3\frac{1}{2}$ years at $2\frac{3}{4}$ per cent. per annum?

4. Find the value of £1·69375.

5. What fraction of an acre is 28 poles?

6. Find by Practice the value of 13 lbs. 7 ozs. 5 dwts. 8 grs. of silver at 3*s*. 9*d*. per oz.

7. Find the least common multiple of 132, 165, 220.

8. A cube of metal, each edge of which measures $\frac{5}{8}$ of an inch, weighs ·625 lb. What is the length of each edge of a cube of the same metal which weighs 40 lbs.?

9. If 9 men do $\frac{3}{5}$ of a piece of work in 14 days, working 10 hours a day, how many extra men must be employed to finish the work in 5 days more if all of them are to work only 8 hours a day?

10. Prove that the difference of the squares of any two odd numbers is divisible by the double of their sum, and that, if the numbers are consecutive odd numbers, the difference of their squares is a multiple of 8. (N.B.—*You may take any two odd numbers which you like to select, show that the propositions are true for those numbers, and then extend your reasoning to all odd numbers.*)

11. A's money is to B's money as 3 : 7; and if B pays A £345, the proportion will be 3 : 5. How much money has each?

12. If 9 men and 6 boys can do in 2 days what 5 men and 7 boys could do in 3 days, in what time could 2 men and 5 boys do the same?

13. A dealer has two sorts of tea, one of which he could sell at 1*s.* 8*d.* per lb. and make 25 per cent. on his outlay, the other at 2*s.* 3*d.* and make 12½ per cent. What profit per cent. will he make if he mixes them in equal quantities and sells the mixture at 1*s.* 11*d.* per lb.?

14. Goods are imported from abroad at an expense equal to 35 per cent. of the cost of production; and the importer makes 15 per cent. on his whole outlay by selling them to a tradesman at £7. 15*s.* 3*d.* per ton. Find the cost per ton of production.

III. ALGEBRA.

(Up to and including the Binomial Theorem, the theory and use of Logarithms.)

[N.B.—*Great importance will be attached to accuracy.*]

1. Divide $x^5+x^4+4x^3+21x^2+23x-40$ by x^2+4x+5.

2. Resolve into factors the expressions:

$$x^3-8y^3, \quad x^2-24xy+128y^2, \quad \text{and} \quad x^4+x^2y^2+y^4.$$

3. Find the Highest Common Factor of the expressions:

$$x^4-8x^3+13x^2-30x+8 \quad \text{and} \quad x^4-4x^3-11x^2-50x+16.$$

4. Prove that $\qquad (b-c)^5+(c-a)^5+(a-b)^5$

is divisible by $\qquad (b-c)(c-a)(a-b),$

and find the other factor.

5. Find the square roots of the expressions

(i.) $n(n+1)(n+2)(n+3)+1$.

(ii.) $24+\sqrt{572}$.

6. Solve the equations

(i.) $\dfrac{x+a}{x-c}+\dfrac{x+c}{x-a}=2$;

(ii.) $x^2+3x+4\sqrt{x^2+3x-3}=48$.

7. If α and β are the roots of the equation $x^2+px+q=0$, prove that $\alpha+\beta=-p$ and $\alpha\beta=q$, and find, in terms of p and q, the equation of which the roots are

$$\alpha+2\beta \quad \text{and} \quad \beta+2\alpha.$$

8. At present the ratio of B's age to A's age is the ratio $5:2$, but in 30 years' time the ratio will be $35:23$; find their ages.

9. Find the least possible positive value of the expression

$$4x+\frac{9}{x},$$

and the greatest possible value of the expression

$$\frac{x+2}{4x^2+16x+25}.$$

10. If the expression $x(1-5x+6x^2)^{-1}$ be expanded in powers of x, prove that the coefficient of x^r is 3^r-2^r.

11. Write down an expression for the number of combinations of n things taken r together, and, if ${}_nC_r$ represent this expression, prove, by general reasoning or otherwise, that

$${}_nC_r={}_nC_{n-r}, \quad \text{and} \quad {}_{n+1}C_r={}_nC_r+{}_nC_{r-1}.$$

12. Having given

$$\log_{10}5=\cdot6989700, \quad \log_{10}7=\cdot8450980, \quad \text{and} \quad \log_{10}11=1\cdot0413927,$$

find the logarithms, to the base 10, of 385 and $\frac{175}{121}$, and solve approximately, to two places of decimals, the equation

$$5^x\cdot7^{x-1}=11^{x+2}.$$

IV. PLANE TRIGONOMETRY AND MENSURATION.

(Including the Solution of Triangles.)

[N.B.—*Great importance will be attached to accuracy.*]

[$\pi = \frac{22}{7}$.]

1. Explain what is meant by the *circular measure* of an angle, and find the circular measure of an angle of x degrees.

A wire AB, 1 foot in length, is bent so as to form an arc of a circle whose diameter is four inches; find the angle subtended at the centre of the circle by the chord AB.

2. Express the Trigonometrical ratios of an angle in terms of the secant, the angle being less than a right angle.

If in a triangle ABC,

$$CA = CB = 2, \quad \text{and} \quad AB = 3;$$

find the value of

$$(\sin A - \cos A)(\sec A + \operatorname{cosec} A).$$

3. Determine the values of the Trigonometrical ratios for an angle of 60°, and for an angle of 30°.

A ladder rests against a vertical wall at an angle of 60° with the horizon, and when the foot is drawn back 18 feet further from the wall the inclination to the horizon is found to be 30°. Find the length of the ladder.

4. Prove that $\cos(90° + A) = -\sin A$ for the case when A is less than two right angles.

Write down the values of $\sin 225°$, $\cos 210°$, $\tan 315°$, and $\operatorname{cosec} 420°$.

5 Show geometrically that $\cos(A + B) = \cos A \cos B - \sin A \sin B$, where A, B are two positive angles whose sum is less than a right angle.

Find the value of $\cos 75°$.

6. Express $\sin 3A$ and $\cos 3A$ in terms of $\sin A$ and $\cos A$, and show that

$$\cos 3A + \sin 3A = (\cos A - \sin A)(1 + 2 \sin 2A).$$

7. Prove the following identities:

(i.) $(x \tan \alpha + y \cot \alpha)(x \cot \alpha + y \tan \alpha) = (x+y)^2 + 4xy \cot^2 2\alpha$.

(ii.) $\cos \beta \cos(2\alpha - \beta) = \cos^2 \alpha - \sin^2(\alpha - \beta)$.

(iii.) $\cos \alpha + \cos 3\alpha + \cos 5\alpha + \cos 7\alpha = \frac{1}{2} \sin 8\alpha \operatorname{cosec} \alpha$.

8. In any triangle ABC find $\tan \frac{1}{2}A$ in terms of the sides a, b, c.

Find the angles of the triangle whose sides are proportional to 3, 5, 7.

9. Show how to solve a triangle when two sides and (i.) an angle opposite to one of the sides, and (ii.) the included angle, are given.

From a boat at sea the angle subtended by the line joining two fixed objects A, B on land is observed to be α; after sailing x yards directly towards A the angle subtended by AB is found to be β, and then after sailing y yards directly towards B, the angle is found to be γ. Find an expression for the distance AB.

10. In a triangle ABC, $a = 35$, $b = 43$, and $C = 75^\circ\ 10'\ 40''$, find the angles A and B.

11. Find the area enclosed by 200 hurdles placed so as to form a regular polygon of 200 sides, the length of each hurdle being 6 feet.

12. A leaden sphere one inch in diameter is beaten out into a circular sheet of uniform thickness = $\frac{1}{100}$th inch. Find the radius of the sheet.

V. STATICS AND DYNAMICS.

[*You may assume that* $g = 32$ *feet per second per second.*]

1. State accurately the principle known as that of the *Triangle of Forces.*

A particle is acted upon by three forces of given magnitudes; show how these forces must be arranged so as if possible to produce equilibrium, when

(i.) the forces have magnitudes represented by the numbers, 6, 11, 18.

(ii.) their magnitudes are represented by 8, 15, 17.

2. Let O be the position of a particle, and OA a right line drawn through O. Find the magnitude and direction of the resultant of forces proportional to 10, 18, 20, 16, acting on the particle, when their lines of action make with OA angles of

$$0°, 30°, 90°, 135°,$$

respectively, all measured in the same sense (*i.e.*, all outwards from O, or all inwards towards O).

3. A body whose mass is 520 pounds is placed on a smooth inclined plane, the tangent of whose inclination to the horizon is $\frac{5}{12}$; find the force necessary to sustain the body

(i.) when this force is horizontal;

(ii.) when it acts along the inclined plane.

4. Define a couple, and show that its moment is the same about all points in its plane.

Prove that the sum of the moments of any two forces in the same plane about any point in the plane is equal to the moment of their resultant about the point.

5. Two parallel forces, P, Q, act on a rigid body; find the magnitude and line of action of their resultant (1) when they act in the same direction; (2) when they act in opposite directions.

Two parallel forces of 20 and 25 pounds' weight, of opposite senses, act on a rigid body, the perpendicular distance between their lines of action being 4 inches; find the resultant.

6. Show how to find the position of the centre of gravity of a given system of particles whose masses are m_1, m_2, m_3, ... occupying given positions.

At each vertex of a triangle is placed a particle whose mass is proportional to the length of the opposite side; show that the centre of gravity is the centre of the inscribed circle.

From a thin uniform circular plate of radius 13 inches is cut out a circular plate of radius 5 inches, the centre of the latter being 4 inches distant from that of the former; find the position of the centre of gravity of the remainder.

7. Describe the screw press, and find the condition for the equilibrium of the effort and resistance ("power" and "weight") applied to it when there is no friction.

8. Define *acceleration.* State clearly what is meant by saying that g is about 32 feet per second per second, and describe any method by which this value of g has been obtained.

Which is the greater acceleration, 15 miles per hour per minute, or $\frac{1}{8}$ feet per second per second?

9. If a particle moves in a right line with constant acceleration, a, having had an initial velocity, u, show that the distance travelled in time t is given by the equation

$$s = ut + \tfrac{1}{2}at^2.$$

A bullet is fired vertically upwards with a velocity of 496 feet per second; 3 seconds afterwards another is fired vertically upwards from the same point with a velocity of 568 feet per second; when and where will they meet?

10. Define an *absolute unit of force*, and, in particular, the *dyne* and the *poundal.*

What force uniformly applied to a mass of 12 pounds will give it an acceleration of 8 feet per second per second? (Express the force both in pounds' weight and in poundals.)

11. From a given point on a horizontal plane is projected a particle with a velocity u at an elevation a; find the range on the plane.

If the particle is projected with a velocity of 500 feet per second at an angle of elevation whose tangent is $\frac{4}{3}$, find the range. Find also the magnitude and direction of the velocity 10 seconds after the time of projection.

FURTHER EXAMINATION.

VI. PURE MATHEMATICS.

[*Full marks may be obtained for about* two-thirds *of this paper.*]

1. Sum up the conditions under which Euclid says that two triangles are equal.

ABC is a triangle and *D* a point in *BC* such that *AD* bisects the angle *A*. If *O* be the centre of a circle which touches *AB* at *A* and also passes through *D*, prove that *OD* and *AC* are at right angles and find the magnitude of the angle *AOD*.

2. Apply the theory of *geometrical progression* to the evaluation of a mixed recurring decimal.

Show that the sum of $\frac{1}{37}$ and the series

$$\frac{38\times 18}{37^2}+\frac{38\times 18^2}{37^3}+\frac{38\times 18^3}{37^4}+\ldots \textit{ ad inf.}$$

is unity.

Employ logarithms to evaluate the tenth and twentieth terms of the series.

3. A clock strikes at intervals of one second. Determine the intervals as they appear to men travelling in express trains at 60 miles an hour directly towards and directly away from the clock respectively. *The velocity of sound may be taken as* 1100 *feet per second.*

4. Find the greatest coefficient in the expansion of

$$(x+y)^n.$$

Prove that the greatest coefficients in the expansion of the trinomial

$$(x+y+z)^{3n+1}$$

have the value

$$\frac{(3n+1)!}{(n!)^3(n+1)}.$$

N.B.—$n!$ represents the product $n(n-1)(n-2)\ldots 3.2.1.$

5. In any triangle show that, in the usual notation,

$$\cos A + \cos B = \frac{a+b}{c}(1 - \cos C),$$

$$\cos B - \cos A = \frac{a-b}{c}(1 + \cos C).$$

6. Find the volume of a pyramid on a triangular base and deduce that the volume of a sphere is the product of the area of its surface and one-third of its radius.

A sphere is cut by a horizontal plane. If P be a point in the perimeter of the section and A the highest point of the sphere, show that the surface of the sphere above the cutting plane has an area equal to that of a circle of radius AP.

7. In the triangle ABC, sides 4, 3, 5 AB produced is a tangent, BC the axis and AC the tangent at the vertex of a parabola.

Draw the triangle in your book and construct geometrically for (i.) the focus, (ii.) the directrix, (iii.) the point P on the parabola at which AB is tangent, (iv.) the other extremity of the focal chord through P, (v.) the extremities of the latus rectum.

8. In an ellipse QQ' is the chord of contact of tangents from an external point P; CK is the perpendicular on QQ' from the centre C; and PG is the perpendicular to QQ', through P meeting the major axis in G. Prove that the semi-minor axis is a mean proportional to KC and PG.

9. Write down the usual forms of equation to a straight line.

Through a given point P whose coordinates are (p, q) a straight line is drawn intersecting the axes in A and B so that P is the middle point of AB. Determine the equation of AB.

10. Find the general equation of the circle when the coordinate axes are inclined to one another at an angle ω.

The axes being rectangular investigate the condition that must be satisfied by the parameters of the circle in order that it may be possible to find a point or points on the circle at equal perpendicular distances from the axes.

11. In the parabola

$$y^2 = 4ax$$

find the equation to the circle passing through the vertex and the extremities of the latus rectum.

Find also the coordinates of the points where the circle is intersected by those normals to the parabola which make an angle of $45°$ with the axis.

12. C is the centre of an ellipse and BB' the minor axis. If S be the focus which (to origin at the centre and coordinate axes coincident with the axes of the ellipse) has a positive abscissa; and $B'S$ be produced to meet the curve in P; show that CP makes an angle ϕ with the major axis such that

$$2e \tan \phi = (1-e^2)^{\frac{3}{2}}.$$

13. Find the equation to the tangent at any point of the hyperbola

$$4xy = a^2+b^2.$$

If two tangents at right angles intersect at P find the locus of P.

VII. MECHANICS.

[*Full marks may be obtained for about* two-thirds *of this paper. Great importance is attached to accuracy.* N.B.—g *may be taken* = 32.]

1. If three forces acting on a particle are in equilibrium, each force is proportional to the sine of the angle between the other two.

$ABCD$ is a quadrilateral, having the angles at A and D right angles, and $CB = CD$. Forces P, Q, and R, acting along AB, CA, and AD respectively, are in equilibrium. If

$$P = \frac{2}{\sqrt{3}} R,$$

show that $AB = \frac{1}{2}BC$ or $\frac{3}{2}BC$.

2. Prove that the algebraical sum of the moments of a number of coplanar forces acting on a particle about any point in their plane is equal to the moment of their resultant about that point.

P is the orthocentre of a triangle ABC. Forces act along AP, BP, CP, and are proportional to $\sin(A+\theta)$, $\sin(B+\theta)$, $\sin(C+\theta)$. Prove that their resultant passes through the centre of the circumscribed circle. What is the value of θ when the forces are in equilibrium?

3. Two equal uniform rods, AB, BC, each of weight W, are hinged together by a smooth hinge at B, and rest on a smooth cylinder of radius b,

whose axis is horizontal. The rods touch the cylinder at points distant one-third of their length from *B*. Find the length of the rods, and prove that the bending moment at the point of contact of either rod with the cylinder is $\frac{2}{3\sqrt{3}} Wb$.

4. A heavy circular wheel is suspended in a vertical plane from a point *B* in a rough vertical wall, by a smooth weightless wire through its centre *O*, and rests at right angles to the wall and touching it at the point *A*. A string is fixed to the rim of the wheel and passes over it on the side away from *A*. If the inclination (α) of the wire *OB* to the horizon is less than 2β, where $\tan\beta$ is the coefficient of friction between the wheel and the wall, show that a force, however great, applied along the string will not turn the wheel unless the direction of the string makes with *BO* produced an angle less than

$$\sin^{-1}\frac{\sin(\alpha-\beta)}{\sin\beta}.$$

(Positive angles are measured from *OA* to *OB*.)

5. Find the position of the centre of gravity of a cylindrical bar with circular ends, whose density at any point is proportional to the square of the distance of the point from the end of the bar.

A cone of vertical angle 2α rests with its base on a rough plane inclined at an angle β to the horizon, the coefficient of friction (μ) being greater than $\tan\beta$. A gradually increasing force is applied at the vertex of the cone parallel to the plane and downwards. Show that, when the force is large enough to disturb equilibrium, the cone will tilt over or slide down according as $\tan\beta$ is less or greater than $\frac{4}{5}(\mu-\tan\alpha)$.

6. A particle initially at rest is acted on by a force constant in direction and magnitude. Prove that the kinetic energy of the particle at any time is equal to the work done on it by the force.

A particle moves from rest down a rough plane inclined at an angle 2θ to the horizon, $\tan\theta$ being the coefficient of friction. Prove that in moving over a length s of the plane it acquires the same velocity as in falling freely through a distance $s\tan\theta$.

7. An engine draws a train whose weight (exclusive of the engine) is 100 tons. The power of the engine is such that when running on the level it exerts a pull of 2 tons weight on the front carriage, and the resistance due to friction, etc., is 11·2 lbs. per ton. Show that if the engine draws the same train from rest up an incline of 1 in 300 it will in one minute acquire a velocity slightly exceeding $15\frac{8}{11}$ miles per hour.

8. A bullet is fired with a velocity of 800 feet per second. Find (*a*) its greatest possible range on the horizontal plane through the point of projection; and (*b*) the height (approximately) to which the bullet ascends when the range is one-tenth of the greatest possible range.

9. A mass of weight W rests on the smooth surface of a horizontal table, also of weight W, and is connected by a weightless string, passing over a smooth pulley at the edge of the table, with a weight $2W$ hanging freely, which is allowed to fall, the string being initially taut. If the table does not move, find the tension of the string, and show that the coefficient of friction between the table and the floor is not less than $\frac{1}{3}$.

10. Two equal imperfectly elastic balls moving in the same straight line impinge directly upon one another. Find the change of kinetic energy produced by the impact.

A billiard ball A moving parallel to one side of the table, strikes another ball B (initially at rest) in the centre. After B has struck the cushion and then struck A again, the velocity of A is three-fourths of its initial velocity, but in the opposite direction. Show that, if e is the coefficient of elasticity between the balls and also between a ball and the cushion,

$$e^3 + e^2 + 3e = 4.$$

MATHEMATICAL EXAMINATION PAPERS

FOR ADMISSION INTO

Royal Military Academy, Woolwich,

NOVEMBER, 1894.

OBLIGATORY EXAMINATION.

I. EUCLID (BOOKS I.—IV. AND VI.).

[*Ordinary abbreviations may be employed, but the method of proof must be geometrical. Proofs other than Euclid's must not violate Euclid's sequence of propositions. Great importance will be attached to accuracy.*]

1. Define *plane rectilineal angle*, *circle*, *gnomon*, *similar rectilineal figures*.

2. If one side of a triangle be produced, the exterior angle is greater than either of the interior opposite angles.

3. The straight lines which join the extremities of two equal and parallel straight lines towards the same parts are also themselves equal and parallel.

ABCD is a quadrilateral; show that, if the four parallelograms *BCDP*, *CDAQ*, *DABR*, *ABCS* be completed, the four straight lines *AP*, *BQ*, *CR*, *DS* will be equal and parallel.

4. If a straight line be divided into any two parts, the square on the whole line is equal to the squares on the two parts, together with twice the rectangle contained by the two parts.

Show also that if a straight line be divided into any three parts, the square on the whole line is equal to the squares on the three parts, together with twice the three rectangles whose sides are the three parts taken two and two together.

5. Describe a square equal to a given rectilineal figure.

6. If a straight line drawn through the centre of a circle bisect a straight line in it which does not pass through the centre, it cuts it at right angles; and if it cut it at right angles, it bisects it.

7. Straight lines in a circle which are equally distant from the centre are equal to one another.

8. If a straight line touch a circle, and from the point of contact a straight line be drawn cutting the circle, the angles which this line makes with the line touching the circle are equal to the angles which are in the alternate segments of the circle.

Four circular coins, of different sizes, are placed upon a table so that each one touches two, and only two, of the remaining three; show that the four points of contact lie on a circle.

9. Inscribe a circle in a given triangle.

Prove that the centre of this circle lies inside each of the three circles described on the three sides of the triangle as diameters.

10. Describe a circle about a given equilateral and equiangular pentagon.

11. Triangles which have one angle of the one equal to one angle of the other and the sides about the equal angles reciprocally proportional, are equal to one another.

Two straight lines AOC, BOD intersect in O and the lines AB, CD are drawn. From the greater of the two triangles AOB, COD cut off a part equal to the less by a straight line drawn through the point O.

12. Parallelograms about the diameter of any parallelogram are similar to the whole parallelogram and to one another.

II. ARITHMETIC.

[N.B.—*The working as well as the answers must be shown.*]

1. Simplify $\frac{2\frac{5}{8} \text{ of } 1\frac{17}{21} - 3}{4\frac{2}{11} - 2\frac{3}{4}} \div 2\frac{3}{11}$.

2. Find the value of (2·37 × ·093) ÷ ·0005.

3. A sum of £237. 18*s.* 4*d.*, lent at simple interest, amounts in three years to £270. 0*s.* 8½*d.* What is the rate per cent.?

4. Find the value in cwts. qrs. and lbs. of ·0234375 of 9⅛ tons.

5. What fraction is 28⅛ cubic inches of a cubic foot?

6. Find the value of 13 acres 2 roods 17½ perches at £14. 14*s.* 8*d.* per acre.

7. Find the greatest common measure of 10058, 4982, and 9823.

8. Justify, from first principles, each step of the process of addition of vulgar fractions and deduce the rule for the addition of decimals.

9. What was the cost of goods on which a man lost 20 per cent. by selling them for £64?

10. A clock set right at noon on Tuesday loses at the rate of 192 seconds in 10 hours. What is the true time on the following Friday afternoon when the reading of this clock is 2 hours 36 minutes?

11. Determine, without performing the divisions, the remainders that result from dividing 48909661 by 8, 16, 25, 9, and 11; and give a brief explanation of the *reason* from which you draw your conclusion in the first three cases.

12. If it costs the same amount to keep 4 horses or 9 oxen, and 5 horses can do as much work as 8 oxen, which will it be more profitable to employ —20 horses and 32 oxen, or 12 horses and 48 oxen?

13. The incomes of two men would be equal if one were increased 7 per cent. and the other diminished $7\frac{1}{2}$ per cent., and the sum of their incomes is £418. 19*s*. What is the income of each?

14. Three trains start from a town *A* at 12.0, 12.5, and 12.10, and travelling each at a uniform rate by 3 different routes of the same length to a town *B*, are observed to pass a signal box at *B* exactly abreast of each other at 12.50. If at 12.20 the sum of the distances traversed by the three trains is 36 miles 7 furlongs, how far from *A* is the signal box at *B*?

III. ALGEBRA.

(Up to and including the Binomial Theorem; the theory and use of Logarithms.)

[N.B.—*Great importance will be attached to accuracy.*]

1. Divide $x^6 - a^6$ by $x^2 - ax + a^2$.

Show that $x^6 - a^6$ is divisible by $x^2 + px + \frac{1}{3}p^2$, if $p^6 + 27a^6 = 0$.

2. Express

$$\frac{\frac{1}{x} + \frac{1}{y-z}}{\frac{1}{x} - \frac{1}{y-z}} \left\{ 1 - \frac{y^2 + z^2 - x^2}{2yz} \right\}$$

in its simplest form.

3. Prove, by the method of finding the Greatest Common Measure, that $2x^2 - x - 6$ is a factor of

$$2x^5 + 3x^4 - 32x - 48,$$

and

$$2x^6 - x^5 - 6x^4 + 2x^3 - 3x^2 - 5x + 6.$$

4. Show, by any method, that

$$a^3(b-c) + b^3(c-a) + c^3(a-b)$$

contains $b-c$, $c-a$, $a-b$ as factors, and find the remaining factor.

5. Find the values of x, y, z from the equations

$$x + 2y - z + 4 = 0,$$

$$3x + 4y + z - 1 = 0,$$

$$5x + 6y - 3z + 18 = 0.$$

6. Simplify the equation

$$\frac{\dfrac{2x-1}{3}-\dfrac{4x^2-1}{x+3}}{x+1-\dfrac{5x^2-9}{3(x-3)}}=\frac{x-3}{x+3}\cdot\frac{10x+1}{2x+3},$$

and solve it.

7. Find the condition that the roots of the equation

$$ax^2+bx+c=0$$

shall be equal.

Determine the values of k for which the equation

$$12(k+2)x^2-12(2k-1)x-38k-11=0$$

will have equal roots.

8. Prove that, if a and l are the first and last terms of an arithmetical progression containing n terms, the sum of the series is

$$\tfrac{1}{2}n(a+l).$$

The sum of 5 terms of an arithmetic series is 10, and the sum of 17 terms is -17; find the series.

9. Expand

$$\frac{1}{\sqrt[3]{1-x}}$$

in a series ascending by powers of x, as far as x^3, by the Binomial Theorem, and write down an expression for the nth term of the series.

10. A and B are two stations on a railway, 90 miles apart. At the same instant one train passes through A towards B and another through B towards A, with different but constant speeds. They pass each other at C, and AC is 10 miles longer than BC; also, the first reaches B half-an-hour before the second reaches A; find their speeds.

11. Define the logarithm of any number to a given base, a.

Find, from a table of common logarithms, the logarithm of 125 to the base $4\frac{1}{2}$. Give, without proof, to four decimal places, the value of the modulus which converts logarithms to base 10 into logarithms to the Napierian base.

12. If e is the Napierian base, prove, by any method, that

$$\log_e(1+x) = x - \frac{x^2}{2} + \frac{x^3}{3} - \frac{x^4}{4} + \dots.$$

Hence show that, for any number, n,

$$\log_e n = 2\left\{\frac{n-1}{n+1} + \frac{1}{3}\left(\frac{n-1}{n+1}\right)^3 + \frac{1}{5}\left(\frac{n-1}{n+1}\right)^5 + \dots\right\}.$$

IV. PLANE TRIGONOMETRY AND MENSURATION.

(Including the Solution of Triangles.)

[N.B.—*Great importance will be attached to accuracy.* $\pi = \frac{22}{7}$.]

1. Prove that the angle subtended at the centre of a circle by an arc equal to the radius of the circle is of the same magnitude for all circles.

Find this angle in degrees.

2. Write down formulæ for $\sin(A+B)$, $\cos(A-B)$, and $\tan(A+B)$, in terms of the trigonometrical functions of A and B.

Find $\cos 75°$ and $\sin 18°$.

3. Find a value of A, positive and less than 90°, satisfying the equation

$$(\sin A° - \cos A°)(\sec A° - \operatorname{cosec} A°) = 2.$$

4. Prove that for all values of A and B

$$\cos A + \cos B = 2\cos\frac{A+B}{2}\cos\frac{A-B}{2}.$$

If the sum of two angles be always equal to a, where a is positive and not greater than 180°, prove that the sum of their cosines will be never greater than $2\cos\frac{a}{2}$ and never less than $2\cos^2\frac{a}{2}$.

5. Prove that for all values of A and B

$$\cos B - \cos A = 2\sin\frac{A-B}{2}\sin\frac{A+B}{2}.$$

Find the numerically smallest value of θ, different from zero, satisfying the equation

$$\cos 3\theta - \cos 4\theta = \cos 5\theta - \cos 6\theta.$$

6. Prove that $\sin 33° + \cos 63° = \cos 3°$;
also that, if $\cot A \cot B = 2$,
then $\cos(A-B) = 3\cos(A+B)$.

7. In the triangle ABC prove that

$$a \cos B + b \cos A = c.$$

$$a \sin B - b \sin A = 0.$$

$$a^2 + b^2 - 2ab \cos C = c^2.$$

8. Given $A = 42^\circ, \quad a = 141, \quad b = 172 \cdot 5,$
find all solutions of the triangle ABC.

9. From the point O the three straight lines OA, OB, OC are drawn in the same plane, of lengths 1, 2, 3 respectively, and with the angles AOB and BOC each equal to 60°. Find the angle ABC correct to one minute.

10. Find the area of the greatest circle which can be cut out of a triangular piece of paper whose sides are 3, 4, 5 feet respectively.

11. A conical extinguisher, whose section through the vertex is an isosceles triangle with vertical angle 30°, is placed over a cylindrical candle whose diameter is one inch, and rests so that the point of contact of the top of the candle with each generating line of the cone bisects that line. Find the whole inside surface of the extinguisher.

V. STATICS AND DYNAMICS.

[*It may be assumed that* $\pi = \frac{22}{7}$, *and that* $g = 32$, *when a foot and a second are the units of length and time.*]

1. Explain why it is that forces can be completely represented by straight lines.

ABCD is a square; find the resultant of the forces represented by the straight lines *AB*, *AC*, and *AD*.

2. Enunciate and prove the theorems of the triangle of forces and the polygon of forces, and state whether the converses of these theorems are true.

3. A heavy pole, weighing 140 lbs., is carried on the shoulders of two men, one at each end; the centre of gravity of the pole being two feet from one end and five feet from the other, find the weight supported by each man.

Also find what would be the effect of placing each man one foot nearer to the centre of gravity of the pole.

4. Find the ratio of the power to the weight when there is equilibrium in a system of three moveable pulleys, each of which is supported by a separate string, and in which the free portions of the strings are vertical.

Also, if the weights of the pulleys, supposed to be equal, are taken into account, find the relation between the power, the weight, and the weight of a pulley.

5. A heavy uniform rod is supported by a string fastened to its ends, of double its own length, which passes over a smooth horizontal rail. Find the tension of the string first, when the rod is hanging at rest in a vertical position, and secondly, when the rod is at rest in a horizontal position.

6. Explain what is meant by saying that a point is moving in a straight line with uniform acceleration, and show how this acceleration is measured.

What is the measure of the acceleration of a body falling freely when eight feet and half a second are the units of length and time?

7. A body is projected vertically upwards with the velocity of 256 feet per second; find the greatest height to which it rises and the time in which it will return to the point of projection.

Also find the times, during the ascent and descent, at which it passes the level of 768 feet above the point of projection.

8. Prove that the path of a projectile is a parabola, and that, if u is the horizontal component of the velocity of projection, the latus rectum of the parabola is equal to $\frac{u^2}{16}$.

9. The top of the spire of a church, standing on a level plane is 200 feet above the plane. From a position on the plane, at the distance of 400 feet from the vertical line through the top of the spire, a bullet is fired off so as to pass horizontally just over the top of the spire. Find the initial direction and the initial velocity of the bullet.

10. Find the direction and magnitude of the acceleration of a point moving uniformly in a circle.

A mass of 7 pounds, on a smooth horizontal plane, is fastened to one end of a string, 7 feet in length, and the other end is fastened to a fixed peg on the plane. The string is then straightened, and the particle is projected horizontally, at right angles to the string, with such a velocity as to describe its circular path in $5\frac{1}{2}$ seconds. Find the tension of the string in poundals, and also in pounds' weight.

FURTHER EXAMINATION.

VI. PURE MATHEMATICS.

[*Full marks may be obtained for about* two-thirds *of this paper.*]

1. Show that the two perpendiculars, erected at the extremities of any chord of a circle, meet any diameter of the circle at two points equidistant from the centre; and contain a rectangle equal to the difference of the squares on the radius and on half the interval they intercept on the diameter.

2. Describe a circle passing through two given points and intercepting, on a given line, a segment of given length.

3. Define a homogeneous integral function of any number of variables.

Write down the most general function, of degree 4, in 3 variables.

4. If the increase in population be 8 per cent. every decade, the rate of increase being constant, find the population of a town of 100,000 inhabitants 5 years hence. Find also, employing a table of logarithms, the percentage of increase per annum.

5. Find the coefficient of xyz in the expansion of

$$\{(1-x-y-z)(1-x-y-z+4xyz)\}^{-\frac{1}{2}}$$

as a rational integral function of x, y, and z.

6. If the circle escribed to the side BC of a triangle ABC touch AB and AC produced in D and E respectively, prove that $AD = AE =$ half the sum of the sides of the triangle.

The longest side of a triangular plot of ground is 100 yards, the perimeter is 250 yards and one angle is 40°. Determine the remaining angles.

7. If

$$r = p(1 + e\cos\phi)^{-1}$$
$$r' = p(1 + e\cos\phi')^{-1}$$
$$r'' = p(1 + e\cos\phi'')^{-1}$$

express

$$\frac{\sin(\phi'' - \phi')}{r} + \frac{\sin(\phi - \phi'')}{r'} + \frac{\sin(\phi' - \phi)}{r''}$$

in a form adapted to logarithmic computation and evaluate it where

$$\phi = 17^\circ\ 4', \quad \phi' = 22^\circ\ 27', \quad \phi'' = 38^\circ\ 19', \quad p = 21.$$

8. There is a rectangular plot of ground. Show how, by means of a cord, an ellipse may be inscribed so as to touch the sides at the middle points.

Prove the propositions on which the construction depends.

9. Give a geometrical construction for drawing tangents to an hyperbola from an external point.

10. Find the equation to a straight line passing through a fixed point (h, k) and making an angle of $\pi/3$ with the axis of x.

If the straight line rotate, in a counter clock-wise direction, about the fixed point through an angle of 1′, show that the intercept on the axis of y is diminished by a length equal to $\frac{\pi}{2700}h$ approximately.

11. If the point (h, k) do not lie on the perimeter of the circle

$$(x - \alpha)^2 + (y - \beta)^2 - \rho^2 = 0$$

interpret the expression

$$(h - \alpha)^2 + (k - \beta)^2 - \rho^2$$

geometrically.

Transpose the above equation to polar coordinates and find the angle between the two tangents from the pole.

12. Find an equation to a parabola.

Prove, analytically, that, at every point of the curve, the diameter and focal radius make equal angles with the tangent.

13. Given the axes of an ellipse, find expressions for

(i.) the distance between the foci.

(ii.) the distance between the directrices.

(iii.) the latus rectum.

(iv.) the product of the lengths of the perpendiculars from the foci on any tangent.

14. Explain carefully the nature of an asymptote to an hyperbola.

Find the equation to a curve, of this description, such that the smaller angle between the asymptotes is 45°, and the distance between the foci 10 units of length.

VII. MECHANICS.

[*Full marks may be obtained for about* two-thirds *of this paper. Great importance is attached to accuracy.* N.B.—*g may be taken* = 32.]

1. Show how to find the resultant of a number of coplanar forces acting at a point.

Forces of magnitudes 3, 4, and 5, act at a point O in directions lying in one plane and making angles of 15°, 60°, and 135° respectively with a line OA in the same plane. Find to two places of decimals the magnitude of the resultant.

2. A small ring of weight W, which can move without friction on a circular wire fixed in a vertical plane, is in equilibrium at a point P on the lower half of the wire under the action of a force R in the direction of the tangent at P to the wire. If the pressure of the ring on the wire is equal to $\frac{1}{2}W$, find the magnitude and direction of the force R.

3. Define the centre of gravity of a rigid body. What assumptions with regard to the action of gravity are made for the purpose of the definition?

$ABCDE$ is a lamina of uniform thickness and density, and of such a shape that $BCDE$ is a square, and $AB = AE$. If the centre of gravity of the lamina is in BE, find the ratio of AB to BC.

4. Find the relation between the power and weight in the wheel and axle.

Show how to arrange three wheels and axles, having radii R and r respectively, so that $P/W = r^3/R^3$.

5. A body of weight 16 lbs. rests on a rough inclined plane inclined at an angle of 30° to the horizon. If a force of 2 lbs. acting up and parallel to the plane is just sufficient to prevent the body from slipping down, find the least force in the same direction which will balance the maximum resistance of the body to motion up the plane.

6. State the proposition known as the "parallelogram of velocities."

Prove that if a point possesses two independent velocities represented by $\lambda \, . \, OA$ and $\mu \, . \, OB$, where OA and OB are two straight lines meeting at O, the resultant velocity will be represented by $(\lambda + \mu)OG$, where G is a point on AB such that $\lambda \, . \, AG = \mu \, . \, GB$.

7. A person rows with a velocity of 6 miles an hour across a river a quarter of a mile wide, which runs with a velocity of 4 miles an hour. The head of the boat makes a constant angle θ with the bank while he rows across, and he arrives at a point 36 yds. 2 ft. lower down the bank than the point opposite his starting point. Prove that $\tan \theta = \frac{4}{3}$.

8. Prove that if a point moves from rest with a constant acceleration f, the distance s passed over in a time t is given by $s = \frac{1}{2} ft^2$.

A body falls from the top of a tower, and after 2 seconds another body is projected downwards with a velocity of 192 feet per second. The two bodies reach the ground at the same time. Find the height of the tower.

9. A balloon when at a height of $2021\frac{1}{2}$ feet from the ground begins to fall with a uniform acceleration of $\frac{1}{16}g$. When the balloon is at a height of $500\frac{1}{2}$ feet from the ground, ballast to the amount of one-tenth the whole mass of the balloon is thrown downwards with a velocity, relative to the balloon, of 10 feet per second. Find the time the ballast will take to reach the ground.

10. A particle is projected from a point P with velocity v in a direction making an angle α with the horizon. Prove that the greatest height above P, to which the particle rises, is $\dfrac{v^2 \sin^2 \alpha}{2g}$.

A stone is thrown from a height of 4 feet so as just to pass horizontally over a wall which is 25 yards distant and 54 feet high. Find the velocity and direction of projection.

MATHEMATICAL EXAMINATION PAPERS

FOR ADMISSION INTO

Royal Military Academy, Woolwich,

JUNE, 1895.

OBLIGATORY EXAMINATION.

I. EUCLID (BOOKS I.—IV. AND VI.).

[*Ordinary abbreviations may be employed, but the method of proof must be geometrical. Proofs other than Euclid's must not violate Euclid's sequence of propositions. Great importance will be attached to accuracy.*]

1. If ABC, DEF be two triangles which have the sides AB, AC equal to the sides DE, DF, each to each, and also the angle ABC equal to the angle DEF; then shall the angles ACB, DFE be either equal or supplementary.

2. Define *parallel straight lines*, *extreme and mean ratio*; and draw some simple figure of a *superficies* which is *not plane*.

If a straight line fall on two parallel straight lines, it makes the alternate angles equal to one another, and the exterior angle equal to the interior and opposite angle on the same side; and also the two interior angles on the same side together equal to two right angles.

3. In any right-angled triangle, the square which is described on the side subtending the right angle is equal to the squares described on the sides which contain the right angle.

4. If a straight line be divided into two equal parts and also into two unequal parts, the rectangle contained by the unequal parts, together with the square on the line between the points of section, is equal to the square on half the line.

5. In obtuse-angled triangles, if a perpendicular be drawn from either of the acute angles to the opposite side produced, the square on the side subtending the obtuse angle is greater than the squares on the sides containing the obtuse angle, by twice the rectangle contained by the side on which, when produced, the perpendicular falls, and the straight line intercepted without the triangle, between the perpendicular and the obtuse angle.

ACB is a straight line, and on *AC* is described an equilateral triangle *DAC*; show that the square on *DB* is equal to the squares on *AC* and *CB*, together with the rectangle contained by *AC* and *CB*.

6. If two circles touch one another externally, the straight line which joins their centres shall pass through the point of contact.

7. In a circle the angle in a semicircle is a right angle; but the angle in a segment greater than a semicircle is less than a right angle; and the angle in a segment less than a semicircle is greater than a right angle.

BEAC is a semicircle whose diameter is *BC*; *D* is any point on *BC*; *AD* is perpendicular to *BDC*; *EB* is equal to *AD*; and *F* is on *DA* produced so that *DF* is equal to *AB*; show that *CE* is equal to *CF*.

8. If from any point without a circle two straight lines be drawn, one of which cuts the circle and the other touches it; the rectangle contained by the whole line which cuts the circle, and the part of it without the circle, shall be equal to the square on the line which touches it.

9. Describe a circle about a given triangle.

10. What is Euclid's method for describing *about* a circle, a regular pentagon, hexagon, or quindecagon?

Show that this method would not apply to the describing about a circle of a triangle equiangular to a given triangle.

11. In a right-angled triangle, if a perpendicular be drawn from the right angle to the base, the triangles on each side of it are similar to the whole triangle, and to one another.

12. Similar triangles are to one another in the duplicate ratio of their homologous sides.

ADOB is the diameter of a circle whose centre is *O*; *C* is a point on the circumference such that *CD* is perpendicular to *AB*; and *EC*, *EA* are tangents to the circle; show that

triangle *ECA* : triangle *OCB* :: *AD* : *DB*.

II. ARITHMETIC.

[N.B.—*The working as well as the answers must be shown.*]

1. Simplify $(2\frac{1}{4}$ of $5\frac{3}{7}) + (3\frac{1}{8}$ of $9\frac{1}{3}) - 10\frac{18}{35}$.

2. Divide 7·777 by 35·35.

3. In what time will the interest on £250, at $3\frac{1}{4}$ per cent. per annum, amount to £60. 18*s.* 9*d.*?

4. Reduce 3 cwt. 3 qrs. 21 lbs. to the decimal of 5 cwt.

5. Find the value of $5\frac{5}{12}$ of £13. 13*s.* 6*d.*

6. What is the rent of a farm of 246 acres, 3 roods, 24 poles, at £2. 5*s.* per acre?

7. Find the greatest common measure of 8775 and 12025.

8. If the water in a tank, 8 feet long and 7 feet wide, is 4 feet deep; and a cubic foot of water weighs 1000 ozs.; what is the weight in tons of the water in the cistern?

9. A certain field could be reaped by 7 men in a certain time, and 5 boys could do as much as 2 men. Find how many boys would be required, in addition to 30 men, for the reaping of a field of twice the size, in a third part of the time.

10. Show that the difference between any improper fraction and unity is always greater than the difference between unity and the reciprocal of the fraction.

[*You may take any improper fraction which you like to select, show that the proposition is true for that fraction; and then extend your reasoning to improper fractions generally.*]

11. The residue of an estate was left to be divided between three persons, *A*, *B*, *C*, in such proportion that *A*'s share was to be to *B*'s share as 4 : 5, and *B*'s share to *C*'s as 9 : 16; the residue realised £2,415. How much was each person entitled to?

12. If you invest a sum of money in such ways that on one-third of it you gain 3 per cent., on one-fifth you lose 4 per cent., and on the remainder gain 6 per cent.; what average rate per cent. do you make on the whole sum invested?

13. Ten years ago a man was three times as old as his son, and five years hence he will be only twice as old. What are their ages respectively?

14. A farmer bought 6 oxen and 100 sheep for £336; of the sheep, 4 died, and the rest were sold at £2. 7*s*. 6*d*. each; and 2 of the oxen fetched £15 each? At what price must the remaining 4 oxen have been sold if the profit on the whole transaction amounted to 5 per cent.?

III. ALGEBRA.

(Up to and including the Binomial Theorem, the theory and use of Logarithms.)

N.B.—*Great importance will be attached to accuracy.*

1. Find the squares of $x+y-2z+1$, and of $x+y-2z-1$. What is the value of the difference of these squares when $z=\frac{1}{2}(x+y)$?

2. Find the L.C.M. of

(1) $(x-1)^2(x^2+2)^3,\ (x-2)^2(x^2-1),\ (x^4-4)$;

(2) $x^5-xy^4,\ x^9+x^8y,\ x^6+y^6+x^2y^2(x^2+y^2)$.

3. Prove the identities

(1) $$\frac{x^3}{(x-y)(x-z)}+\frac{y^3}{(y-z)(y-x)}+\frac{z^3}{(z-x)(z-y)}=x+y+z;$$

(2) $$\frac{a^3-y^3}{a^4-y^4}-\frac{a-y}{a^2-y^2}-\frac{1}{2}\left\{\frac{a+y}{a^2+y^2}-\frac{1}{a+y}\right\}=0.$$

4. Solve the equations

(1) $x^2-17x+72=0$.

(2) $\begin{cases} 11x-13y=40 \\ 10x-12y=35+\frac{9}{11}. \end{cases}$

(3) $\begin{cases} 3x^2+4xy+5y^2=\frac{71}{36} \\ 5x^2+4xy+3y^2=\frac{9}{4}. \end{cases}$

5. *A* and *B* have the same birthday. *A*'s age is represented, on his birthday in the year 1895, by the two right-hand digits of the year in which *B* was born; also the product of *A*'s age and the number represented by the two right-hand digits of the year in which *he* was born gives the year when *B* was 9 years old. Find their ages, *A* being older than *B*.

6. Find the sum of an infinite geometrical progression whose first term is a and ratio r.

To what restriction is r subject, and why?

During any year the excess of births over deaths causes an increase of h per cent. of the population, and at the end of the year a fixed number A of people emigrate. Prove that at the end of n years a population P becomes

$$b^n P - \frac{b^n - 1}{b - 1} A,$$

where
$$b = 1 + \frac{h}{100}.$$

7. Investigate the number of different ways in which n men may stand in a row.

If two specified men are, neither of them, to be at either extremity of the row, show that the number of arrangements is

$$(n-2)(n-3) \times (n-2)!$$

8. Write down the $(s+1)^{\text{th}}$ term in the expansion of

$$(a+b)^n.$$

Find the sum of all the coefficients; and show, by using a table of logarithms that if $a+b=1$ and $n=10$, the first *term* is greater than the sum of all the remainder if a is greater than (about) ·933.

9. Eliminate a from the equations

$$x = \log_a b, \quad y = \log_a c.$$

Prove that

$$z = 1 + \frac{\log z}{1!} + \frac{(\log z)^2}{2!} + \frac{(\log z)^3}{3!} + \ldots.$$

Employ tables to find the square root of π to 5 places of decimals.

10. If

$$y^6 + y^5 - 5y^4 - 4y^3 + 6y^2 + 3y - 1 = 0,$$

and
$$y = x + \frac{1}{x},$$

show that
$$\frac{x^{13} - 1}{x - 1} = 0.$$

IV. PLANE TRIGONOMETRY AND MENSURATION.

(Including the Solution of Triangles.)

[N.B.—*Great importance will be attached to accuracy.*]

1. Explain the measurement of angles in *circular measure.*

If an angle contains A seconds and its circular measure is a, show that, approximately,

$$A'' = 206265 \times a.$$

Find from the tables the values of $\cos\left(\frac{5}{8}\right)$ and $\sec\left(\tan\frac{\pi}{6}\right)$ disregarding seconds.

2. Show how to construct the angle whose cotangent is $\frac{8}{15}$, and find—

(*a*) The sine of this angle;
(*b*) The cosine of its half.

3. If x and y are any two numbers, show that the equation

$$\sin\theta = \frac{2xy}{x^2+y^2}$$

always gives a real value of the angle.

With this value for $\sin\theta$, find the value of $\tan\frac{\theta}{2}$.

4. Prove that

$$\sin A + \sin B = 2\sin\frac{A+B}{2}\cos\frac{A-B}{2}.$$

Prove also that

$$\sin(A-B)+\sin(B-C)+\sin(C-A) = -4\sin\frac{A-B}{2}\sin\frac{B-C}{2}\sin\frac{C-A}{2},$$

assuming the necessary elementary formulæ.

5. Prove the formula

$$\tan(A+B) = \frac{\tan A + \tan B}{1 - \tan A \tan B}.$$

Show that $\quad \cos^{-1}\frac{8}{17} - \tan^{-1}\frac{4}{3} = \tan^{-1}\frac{13}{84}.$

6. Prove the formulæ

$$\cos 2A = 2\cos^2 A - 1,$$
$$\cos 3A = 4\cos^3 A - 3\cos A.$$

7. Show that the sines of the angles of a plane triangle are proportional to the lengths of the opposite sides, and deduce the relation

$$\cos\frac{A-B}{2} = \frac{a+b}{c}\sin\frac{C}{2}.$$

8. ABC is a plane triangle, and P a point in the side AB such that

$$\frac{AP}{BP} = \frac{m}{n}.$$

If the angle CPB is θ, show that

$$(m+n)\cot\theta = n\cot A - m\cot B.$$

If the angles ACP and BCP are α and β respectively, show also that

$$(m+n)\cot\theta = m\cot\alpha - n\cot\beta.$$

9. Given $b = 14$, $c = 13$, and $A = 67^\circ\ 22'\ 48''$ in a triangle, find C by logarithmic calculation.

10. The radii of the circular faces of a frustum of a right cone are 12 and 8 feet, and the area of its curved surface is $20\pi\sqrt{241}$ square feet; find the thickness of the frustum.

Show that the vertical angle of the cone, of which this is a frustum, is $29^\circ\ 51'\ 46''$.

V. STATICS AND DYNAMICS.

[*Assume that* $\pi = \frac{22}{7}$, *and that* $g = 32$.]

1. If a number of forces, lying in one plane, act at a point, explain how their resultant may be found.

A, *B*, *C*, *D* are the angular points of a square taken in order, and forces represented in direction by the lines *AB*, *BD*, *DA* and *AC*, and in magnitude by the numbers 1, $2\sqrt{2}$, 3 and $\sqrt{2}$, act at a point; find their resultant graphically or otherwise.

2. Prove that two forces, whose lines of action intersect, have moments about a point in their plane, that are together equal to the moment of their resultant about the same point.

P is a fixed point on the circumference of a fixed circle; *PM* and *PN* are any two chords of the circle at right angles to one another; *Q* is any other fixed point whatever in the plane of the circle. Show that if *PM* and *PN* represent forces, the algebraic sum of their moments about *Q* is constant.

3. Define the centre of gravity of a body; and show that a body cannot possess more than one centre of gravity.

Prove that the centre of gravity of three particles of equal mass placed at the angular points of a triangular lamina of uniform thickness coincides with the centre of gravity of the lamina.

4. A bent lever consists of two uniform, heavy, straight rods, whose lengths are as 3 to 4; find the weight which must be attached to the end of the shorter rod in order that—the fulcrum being at the junction of the two rods—they may make equal angles with the horizon.

5. Find the relation between the Effort or "Power" and the Resistance in the case of the frictionless screw press, when motion is just about to take place.

The step—or distance between two threads—of a screw is 0·187 of an inch, the length of the arm (reckoned from the centre of the screw) on which the effort acts is 25 inches, and the effort is 11·9 lbs.; find the resistance when the screw is on the point of moving.

6. Express, in feet per second, the difference between a velocity of 60 yards per hour and a velocity of $3\frac{1}{2}$ feet per minute.

What would be the average velocity of a body which went 30 yards at the first rate, and then 35 yards at the second rate?

7. Find the acceleration of a particle, which is in a state of uniformly varying motion (1) if the velocity increases from 5 feet per second to 8 feet per second, while the particle describes a space of 13 feet; (2) if the spaces described in the first and sixth seconds are 7 and 17 feet respectively.

8. A body slides from rest down a smooth inclined plane of length 192 feet and height 12 feet; find

(i.) the acceleration of the body while sliding;

(ii.) the velocity acquired in sliding from the top to the bottom of the plane;

(iii.) the time taken to get from the top to the bottom of the plane.

9. What is the meaning of uniform angular velocity? Find in radians per second the angular velocity of the minute hand of a clock, keeping correct time.

If T is the periodic time of a particle revolving uniformly in a circle of radius r, show that the acceleration of the particle directed towards the centre is $\frac{4\pi^2}{T^2}r$.

10. Show that a heavy body projected obliquely in vacuo will describe a parabola.

Find the direction of projection when the range in a horizontal plane is $4\sqrt{3}$ times the greatest height.

FURTHER EXAMINATION.

VI. PURE MATHEMATICS.

[*Full marks may be obtained for about* two-thirds *of this paper.*]

1. Solve the equations

$$\text{(i.)}\quad (a-1)(x+1)(x+a^3)=x(a^4-1).$$

$$\text{(ii.)}\quad \left.\begin{aligned}\frac{x}{y+a}+\frac{y}{x+a}&=1\\ x^2+y^2&=b^2\end{aligned}\right\}.$$

2. The interest on a sum of p pounds for a certain time is i pounds, and the discount at the same rate of interest for the same time is d pounds. Show that

$$\frac{1}{p}=\frac{1}{d}-\frac{1}{i}.$$

3. The index n being a positive integer, show that

(i.) All the coefficients in the expansion of $(1+x)^n$ are integers;

(ii.) The coefficients of terms equidistant from the beginning and end of the same expansion are equal;

(iii.) The sum of the even coefficients equals the sum of the odd coefficients.

4. If $u_n=\sin^n\theta+\cos^n\theta$, prove that

$$\frac{u_3-u_5}{u_1}=\frac{u_5-u_7}{u_3}.$$

5. A person travelling uniformly at the rate of 45 miles per hour along a straight line of railway observes the altitude of the top of a distant steeple at intervals of ten seconds. If two consecutive observations be 11° 41′ and 18° 35′ respectively, find what must be the height of the steeple that the next observation may be identical with the first of the former two.

6. *AHK* is an equilateral triangle, and *ABCD* is a rhombus whose sides are equal to the sides of the triangle, and *BC*, *CD* pass through *H*, *K* respectively. Prove that the angle *A* of the rhombus is ten-ninths of a right angle.

7. Lines are drawn joining the angular points *A*, *B*, *C* of a triangle to any point *O* in its plane. Prove that the lines from the middle points of the sides *BC*, *CA*, *AB* respectively parallel to *OA*, *OB*, *OC*, meet in a point.

8. Show that the straight line joining the points (24, 16) and (−21, −14) passes through the origin, and determine the co-ordinates of the points of trisection of this line.

9. Explain how to find the length of the perpendicular from the point (h, k) on the line $x \cos \alpha + y \sin \alpha = p$.

Find the equations of the lines bisecting the angles between the straight lines

$$3x + 4y = 5, \qquad 12x - 5y = 41.$$

10. Prove that

$$x^2 + y^2 - 4x - 2y + 4 = 0$$

represents a circle, and find the length of its radius.

Show that the lines $x = 1$, $y = 2$ each touch the circle, and find the other co-ordinates of the points of contact.

11. Find the equation to a tangent to the ellipse

$$\frac{x^2}{a^2} + \frac{y^2}{b^2} = 1$$

in terms of the eccentric angle of the point of contact.

Prove that two tangents to an ellipse which are at right angles to each other intersect on a fixed circle concentric with the ellipse.

12. Find the equation to the tangent at any point of the rectangular hyperbola $xy = c^2$.

If $c \tan \theta$, $c \cot \theta$ be the co-ordinates of a point on the curve, show that the chord through the points θ and ϕ, where $\theta + \phi$ is constant, passes through a fixed point on the conjugate axis of the hyperbola.

13. Given a chord of a parabola and the direction of the axis, show that the locus of the focus is a hyperbola whose foci are at the extremities of the given chord.

14. Through one of the vertices A, and the extremities P, P', of a double ordinate of an ellipse or hyperbola, a circle is drawn cutting the axis again in K. If G be the foot of the normal at P, prove that GK is of constant length.

VII. MECHANICS.

[*Full marks may be obtained for about* two-thirds *of this paper. Great importance is attached to accuracy.* N.B.—*g may be taken* = 32.]

1. Find the magnitude of the resultant of two forces P, Q which act at a point, the angle between their directions being θ.

Find the resultant of two forces 3 lbs. and 5 lbs. which act at an angle of 60°; and show that its magnitude will be unaltered if either of the given forces be replaced by a force of 8 lbs. acting in the opposite direction.

2. State the necessary and sufficient conditions for the equilibrium of three parallel forces acting upon a rigid body.

A bookshelf supported at its extremities is just filled by two sets of books, the books of each set being placed together. One set consists of 14 volumes, each $1\frac{1}{2}$ inches thick and weighing $2\frac{1}{2}$ lbs.; the other consists of 12 volumes, each $1\frac{1}{4}$ inches thick and weighing 2 lbs. Find the pressures on the supports, the weight of the shelf being 8 lbs.

3. The weight and centre of gravity of a body, and also of a portion of the body, being known, show how to determine the centre of gravity of the remainder.

A figure is formed by taking away from a square the triangle whose angular points are the middle points of three of the sides. Find the position of its centre of gravity.

4. State the laws of Limiting Friction, and explain what is meant by the "Coefficient of Friction."

A uniform beam AB whose length is $12\frac{1}{2}$ feet rests with one extremity A on a rough horizontal plane AC and is kept from falling forwards by a cord BC, 20 feet long, whose extremity is attached to a fixed point C in the plane, directly behind the beam. If the beam be on the point of slipping when $AC = AB$, find the coefficient of friction.

5. Find the relation between the power and the weight in that system of pulleys in which all the strings are attached to the weight, the weights of the pulleys being equal.

If there be one fixed, and two moveable, pulleys, find how far the weight can be practically raised, if it be initially 24 feet below the lower moveable pulley.

6. Explain how velocity is measured, and if u be the measure of a velocity when s feet and t seconds are the units of space and time, find its measure when the units are s' feet and t' seconds.

Compare the velocities of two particles, one of which describes 9 miles in two hours and the other 11 feet in 4 seconds.

7. Express the space passed over by a particle moving subject to uniform acceleration in terms of its initial and final velocities and the time occupied.

An engine-driver reduces the speed of a train (at a uniform rate) from 40 to 30 miles per hour in a quarter of a minute. Find the distance passed over in this time, and also the velocity of the train, when half this distance has been described.

8. A ball of mass m impinges directly upon a ball at rest of the same size but of mass m'. Show that after impact the balls will move in the same direction or in opposite directions according as m is $>$ or $< em'$, e being the coefficient of elasticity.

9. A particle is projected from O with velocity u in a direction inclined to the horizon at an angle α. Prove that the equation of its path is

$$y = x \tan \alpha - \frac{gx^2}{2u^2}(1 + \tan^2 \alpha),$$

the axes of x and y being the horizontal and vertical lines drawn in the plane of the motion through the point of projection.

Find the velocity and angle of projection of a particle which being thrown from the level of the ground just clears a wall 18 feet high at a distance of 36 feet from the point of projection, and strikes a wall parallel to the former and 60 feet beyond it, at a point 8 feet above the ground. The plane of projection is perpendicular to the walls.

10. A particle of weight W, attached by a string of length L to the vertex of a smooth cone whose axis is vertical and semi-vertical angle α describes a horizontal circle on the surface of the cone with uniform velocity v; find the tension of the string and the pressure on the surface.

MATHEMATICAL EXAMINATION PAPERS

FOR ADMISSION INTO

Royal Military Academy, Woolwich,

NOVEMBER, 1895.

OBLIGATORY EXAMINATION.

I. EUCLID (BOOKS I.—IV. AND VI.).

[*Ordinary abbreviations may be employed, but the method of proof must be geometrical. Proofs other than Euclid's must not violate Euclid's sequence of propositions. Great importance will be attached to accuracy.*]

1. If two triangles ABC, DEF, have the sides AB, AC of the one respectively equal to the sides DE, DF of the other, and the angle BAC equal to the angle EDF, the triangles are equal in all respects.

2. Prove that triangles on equal bases and between the same parallels are equal to one another.

If two triangles have equal bases, but the height of one be double the height of the other, prove, by Euclid's methods, that one of the triangles is double the other.

3. If the square on one side of a triangle be equal to the squares on the other two sides together, prove that these sides include a right angle.

4. If a straight line AB be divided internally at any point C, prove that the square on AB is greater than the squares on AC, CB together, by twice the rectangle contained by AC and CB.

If the base BC of a triangle ABC be bisected at D, prove that the squares on AB, AC are together equal to twice the squares on BD and DA together. Show that this includes some of Euclid's propositions, in Book II., as particular cases.

5. Show how to divide a given straight line into two parts, so that the square on one part may be equal to the rectangle contained by the whole line and the other part.

6. Define a tangent to a circle, and prove that it is at right angles to the diameter of the circle through its point of contact.

7. Show that any two opposite angles of a quadrilateral inscribed in a circle are together equal to two right angles.

Show that two opposite sides of a convex quadrilateral (*i.e.*, a quadrilateral without re-entrant angles) described about a circle are together equal to the other two sides together.

State a sufficient condition that it may be possible to inscribe a circle in a given convex quadrilateral, proving your result.

8. Through a point O interior to a circle two chords AOB, COD, are drawn; prove that the rectangle AO, OB is equal to the rectangle CO, OD.

9. Describe a circle interior to a given triangle to touch the sides of the triangle.

Show that four circles can be drawn to touch the sides of a triangle, three of them being exterior to the triangle.

10. Describe an isosceles triangle having each of the angles at the base double of the third angle.

11. If the vertical angle BAC of a triangle be bisected internally by a line cutting the base in D, prove that the ratio $BD : DC$ is equal to the ratio $BA : AC$.

If the perpendicular from *C*, upon the bisector *AD*, meet *AD* in *N*, and *O* be the middle point of the base *BC*, prove that *ON* is half the difference of the sides *AB*, *AC*.

12. If in the triangles *ABC*, *DEF*, the angles *BAC*, *EDF* be equal, and the ratio *BA* : *ED* be equal to the ratio *FD* : *CA*, prove that the triangles are equal.

II. ARITHMETIC.

[N.B.—*The working as well as the answers must be shown.*]

1. Simplify

$$\frac{5\frac{1}{3} \text{ of } 3\frac{1}{3} \text{ of } 2\frac{1}{7} - 6\frac{1}{4} \text{ of } 3\frac{1}{5} \text{ of } 1\frac{1}{8}}{7\frac{3}{5} \text{ of } 3\frac{4}{7}}.$$

2. Divide 3·425 by ·002192.

3. What principal, if invested for 3 years at $2\frac{3}{4}$ per cent. per annum simple interest, will amount to £575. 10*s*. 7*d*.?

4. Find the value of

$2\frac{1}{7}$ of $3\frac{1}{8}$ of 4 lbs. 8 oz. 10 dwt. 12 grs. Troy.

5. Reduce £4. 18*s*. $10\frac{1}{2}$*d*. to the decimal of 5 guineas.

6. Find the cost of 276 tons 16 cwt. at £3. 18*s*. $11\frac{1}{2}$*d*. per ton.

7. Find the least common multiple of 385, 231, 165, 105.

8. A room, 21 ft. 4 in. long, 18 ft. 8 in. wide, and 15 ft. 6 in. high, is papered with paper 32 inches wide at one shilling a yard. What is the total cost?

9. If by selling a certain horse for £66 I should lose 28 per cent. of the cost at which I bought the animal, what is my loss?

10. Prove that the product of any two numbers which consist of three figures and four figures respectively must be a number consisting of not less than six nor more than seven figures.

11. A cubical box of external dimensions 17 inches each way would contain crushed ore of the value of £421. 17*s*. 6*d*. if it were made of material 1 inch thick; but by mistake it has been made of thicker material, and the difference in the value of the ore which it will hold is consequently £78. 17*s*. 6*d*.; what is the real thickness of the material?

12. Three persons contribute sums of £250, £500, and £750 respectively, towards a venture, on the understanding that the profits shall be divided in such a way that the *rate* of interest which each receives shall be in proportion to the amount of his contribution. If the profits for a year amount to £245, how much will each of them receive?

13. A train, going at the rate of 72 miles an hour, overtakes another train 192 yards long, going in the same direction on a parallel line at the rate of 54 miles an hour, and completely passes it in three-fourths of a minute: find the time in which the trains would have completely passed one another, if they had been going in opposite directions, and the length of the faster train.

14. Supposing the quantity of land under barley in England to be this year half as much again as that under wheat, and the quantity under oats to be equal to the other two together; if the quantity under wheat next year be reduced by 25 per cent., and the quantity under barley increased by 5 per cent., the whole quantity remaining the same as before, by how much per cent. will the quantity under oats be increased?

WOOLWICH ENTRANCE EXAMINATION.

III. ALGEBRA.

(Up to and including the Binomial Theorem, and the theory and use of Logarithms.)

[N.B.—*Great importance will be attached to accuracy.*]

1. **Divide** $x^5+x^4-9x^3+33x^2-7x-49$ **by** x^2+3x-7.

2. **Find the factors of**

(i.) $x^2+16x+63$.

(ii.) $y^3-43a^2y+42a^3$.

(iii.) $x^7-14x^5+49x^3-36x$.

3. **Find the highest common factor of** $x^4-3x^3-5x^2-18x+4$ **and** $4x^4-18x^3-5x^2-3x+1$.

4. **Prove that** $a+b+c$ **is a factor of** $a^4+b^4+c^4-2b^2c^2-2c^2a^2-2a^2b^2$, **and find the other factors.**

5. **Solve the equations**

(i.) $x^2-11x+30=0$;

(ii.) $\left.\begin{array}{l} 4x+7y=-1 \\ 3x-y=3 \end{array}\right\}$;

(iii.) $\left.\begin{array}{l} x^2+y^2=5 \\ x^2-y^2=\frac{3}{2}xy \end{array}\right\}$.

6. **A man leaves half his property to his eldest son, three-quarters of the remainder to his second son, and four-fifths of what is then left to his third son. If there is still £100 not disposed of, find the amount of the whole property.**

7. **An express train is timed to run at a uniform speed from a point** ***A*** **to a point** ***C*****.** ***B*** **is a point on the line such that** ***AB*** **is three-fifths of the whole distance** ***AC*****. In running from** ***A*** **to** ***B*** **the train has an average speed** $1\frac{1}{2}$ **miles an hour below its normal amount, and is consequently two minutes late at** ***B*****. The driver arrives punctually at** ***C*** **by running from** ***B*** **to** ***C*** **at an average speed** $2\frac{2}{7}$ **miles an hour above the normal amount. Find the normal speed, and the distance from** ***A*** **to** ***C*****.**

8. If a, b, c, d are consecutive terms of a geometrical progression show that $a^2+(b+c)(b+d)=c^2+(a+c)(a+d)$.

The first term of a geometrical progression is 1, and the common ratio is $\frac{2}{3}$. Find the sum of the first six terms; and find also (by using logarithmic tables) the number n of terms which must be taken in order that the difference between the sum of the first n terms and the sum to infinity may be less than $\frac{1}{2\times 10^7}$.

9. Show that the least possible value of the expression $x^2+2px+q$, for real values of x, is $q-p^2$.

If α, β are the roots of the equation $x^2+2px+q=0$, find the value of $(\alpha-\beta)^2$.

10. There are six gentlemen and nine ladies at a lawn-tennis party, and two courts are available. In how many ways can two "double" sets (*i.e.*, two sets of four players each) be made up, each pair of players consisting of a lady and a gentleman; the particular courts and sides taken by the different pairs not being taken into account?

11. Write down the first six *coefficients* in the expansion of $(1-2x)^{-\frac{1}{2}}$.

Find the greatest *term* in the expansion of $(2+3x)^7$ when $x=\frac{4}{5}$.

12. Prove that $\log_a xy = \log_a x + \log_a y$.

Find, by using logarithmic tables, the value of

$$\frac{\sqrt[3]{100}}{\sqrt[2]{123}}$$

to six places of decimals.

IV. PLANE TRIGONOMETRY AND MENSURATION.

(Including the Solution of Triangles.)

[N.B.—*Great importance will be attached to accuracy.*]

1. Write down the relation which exists between the measures of an angle in degrees and in radians.

What is the measure (i.) in degrees (ii.) in radians of an internal angle of a regular decagon?

2. Prove geometrically the formula

$$\sin 2A = 2 \sin A \cos A, \text{ when } A \text{ is} < 45^\circ.$$

Express $1 - \sin^6 A - \cos^6 A$ in terms of $\sin 2A$.

3. Obtain the value of $\sin 54^\circ$.

If $\sin 4\theta \cos \theta = \frac{1}{4} + \sin \frac{5\theta}{2} \cos \frac{5\theta}{2}$, find *one* value of θ.

4. Find an expression for all the angles which have a given tangent.

Find the general value of x that will satisfy the equation

$$\tan x - \sqrt{3} \cot x + 1 = \sqrt{3}.$$

5. Establish the identities

(i.) $(\operatorname{cosec} A - \sin A)(\sec A - \cos A) = (\tan A + \cot A)^{-1}$.

(ii.) $\dfrac{\tan \theta}{(1 + \tan^2 \theta)^2} + \dfrac{\cot \theta}{(1 + \cot^2 \theta)^2} = \frac{1}{2} \sin 2\theta$.

(iii.) $\sin^{-1} \frac{4}{5} + \sin^{-1} \frac{5}{13} = \sin^{-1} \frac{63}{65}$.

6. If $A + B + C = 180^\circ$, prove that

$$\cos A + \cos B + \cos C - 1 = 4 \sin \frac{A}{2} \sin \frac{B}{2} \sin \frac{C}{2};$$

and if $A + B + C = 90^\circ$, prove that

$$\sin 2A + \sin 2B + \sin 2C = 4 \cos A \cos B \cos C.$$

7. Find an expression for the cosine of half of an angle of a triangle in terms of the sides.

Prove that in any triangle

$$\frac{a}{bc}+\frac{\cos A}{a}=\frac{b}{ca}+\frac{\cos B}{b}=\frac{c}{ab}+\frac{\cos C}{c}.$$

8. In any triangle if

$$\tan\phi=\frac{a-b}{a+b}\cot\tfrac{1}{2}C,$$

prove that (without regard to sign)

$$c=(a+b)\frac{\sin\frac{1}{2}C}{\cos\phi}.$$

9. The sides of a triangle are 237 and 158, and the contained angle is 58° 40′ 3·9″. Find the value of the base, without previously determining the other angles.

10. If r and R are the radii respectively of the circles inscribed in, and described about the triangle ABC, show that

$$r=4R\sin\frac{A}{2}\sin\frac{B}{2}\sin\frac{C}{2}.$$

11. Three halfpennies are placed on a flat table in contact with one another, and with their centres forming an equilateral triangle.

Find the area of the space enclosed between them. If a fine string is wound tightly round them so that each of the free portions of the string is a tangent to two of the coins, what is the length of the string?

Diameter of a halfpenny = one inch. $\pi=3{\cdot}1416$, $\sqrt{3}=1{\cdot}73205$.

V. STATICS AND DYNAMICS.

[*g may be taken* = 32.]

1. Show that forces may be represented by straight lines.

If a straight line AB represent a force of 1 lb.; construct the line which shall represent a force of $3\sqrt{2}$ lbs.

Three forces of 1, 1 and $\sqrt{2}$ lbs., acting at a point, are in equilibrium; find, graphically or otherwise, the angle between the greatest force and either of the other forces.

2. State the necessary and sufficient conditions that three parallel forces acting upon a rigid body may be in equilibrium.

A heavy uniform bar $ACDB$ rests in a horizontal position upon two fixed supports C, D, whose distance apart is 6 inches and equal to the length of the projecting part AC of the bar. If an upward force of 2 lbs. applied at A just lifts the bar off the support C, and a downward force of 8 lbs. at A justs lifts it off D, find the length and weight of the bar.

3. Three forces, which are not parallel, act in one plane upon a rigid body and keep it at rest; prove that their lines of action meet in a point.

A square lamina $ABCD$, whose weight is 4 lbs., can turn in a vertical plane about a hinge at A. Find the force which, acting along BC, will keep the square in a position with this side horizontal; find also the magnitude and direction of the hinge action at A.

4. Show how to determine the centre of gravity of a system of heavy particles lying in one plane.

Four equilateral triangular laminæ, each of side a, but of different weights, $3W$, $5W$, $5W$, $7W$, are placed with their sides in contact and with the heaviest triangle in the middle so as to form an equilateral triangle of side $2a$. Find the position of its centre of gravity.

5. Draw any system of pulleys with parallel strings by means of which a force may balance a weight seven times as great.

If the direction of the force be vertical, find through what distance its point of application must move in order to raise the weight through 5 feet.

6. Enunciate and prove the proposition known as the "Parallelogram of Velocities."

From a ship sailing in a north-easterly direction at 15 miles an hour it is observed that a second ship is always south of it; supposing that this second ship is sailing eastwards, find the rate at which it is travelling.

7. A particle starts from rest subject to a given uniform acceleration a. Write down the formulæ connecting s, the space described, v, the velocity acquired, and t, the time during which the motion takes place.

What will these formulæ become if the particle *starts* with velocity v subject to a *retardation* a?

8. State the usual relation between the measure of the mass of a body and that of the force which produces in it an acceleration a. Express the unit of force in terms of the units of mass, space, and time.

At A, B masses of 2 and 3 lbs. are respectively placed, and each mass is acted upon by a force equal to the weight of the other, and in a direction from it towards the other. If they start from rest at the same moment, and meet at the end of 3 seconds, find the distance AB and the velocities of the masses at the moment of striking.

9. Two bodies of given masses are suspended at the ends of a string which passes round a smooth pulley without weight. Find the acceleration, and the tension of the string.

If the masses be 8 lbs. and 17 lbs., and the former be initially 4 feet below the latter, find after what interval the vertical distance between the bodies will be again 4 feet.

10. A particle is projected with a given velocity and in a given direction; find the range on a horizontal plane through the point of projection.

If the particle pass through a vertical plate at the highest point of its path and have its velocity diminished in consequence by one-third, find the corresponding diminution in the range.

FURTHER EXAMINATION.

VI. PURE MATHEMATICS.

[*Full marks may be obtained for about* two-thirds *of this paper.*]

1. What is the present value of an annuity of £400 per annum for 20 years, beginning one year from the present date, allowing compound interest at 4 per cent. per annum?

2. Two cyclists start together to race from Cambridge to Saffron Walden and back. The faster rider, who maintains a speed of fifteen miles per hour, is $2\frac{3}{4}$ miles ahead of the other on reaching Walden. After covering six miles of the return journey, the tyre of the leader is punctured and occupies 25 minutes to repair. The other rider passes him 7 minutes before he is ready to start again. Find who wins the race, and at what pace the slower rider travels.

3. Show how to find the greatest term in the expansion of $(a+x)^n$ by the Binomial Theorem.

Employ the Binomial Theorem to find the values of $\cdot 9^7$ and $\cdot 99^4$ to five places of decimals.

4. If α, β are unequal values of θ satisfying the equation

$$a \tan\theta + b \sec\theta = 1,$$

find a and b in terms of α and β, and prove that

$$\sin\alpha + \cos\alpha + \sin\beta + \cos\beta = \frac{2b(1-a)}{1+a^2}.$$

5. ABC is a triangle, and D, E, F, are the middle points of BC, CA, AB respectively. If the lengths p, q, r of the lines AD, BE, CF are given, solve the triangle.

If the measurements of p, q, r are slightly incorrect, and the small errors be x/p, y/q, z/r, find the consequent error in the computed value of the angle A when $y = z = -2x$, and show that the calculated values of b and c are not affected by these errors.

6. O is a fixed point on a circle, and PQ a chord of the circle such that the sum of the squares on OP and OQ is constant. Show that the middle point of PQ lies on a fixed straight line.

7. Prove that the perpendiculars from the angular points of a triangle on the opposite sides intersect in a point.

This point being called the orthocentre, let P, Q, R, S be the orthocentres of the triangles BCD, CDA, DAB, ABC, where A, B, C, D lie on one and the same circle; prove that the quadrilaterals $ABCD$, $PQRS$ are of the same size and shape.

8. What points are represented by the equations

$$x^2+y^2=13, \quad xy=6?$$

Show their relative positions on a figure, and demonstrate that the four points are the corners of a parallelogram.

9. Find the general equation to a straight line which passes through the point of intersection of the two given straight lines

$$ax+by=1, \quad lx+my=1.$$

Determine the equation of the straight line which passes through the origin and the point of intersection of the lines

$$5x-3y=11, \quad x+2y=10.$$

10. Find the equation of the circle which has its centre at the point (9, 4) and passes through the point (1, −2), and show that it also passes through the point (3, 12).

Find the equation to the straight line which is the polar of the origin with respect to this circle.

11. Show that the line

$$y=mx+\frac{a}{m}$$

always touches the parabola $y^2=4ax$.

Show that the middle points of chords parallel to this tangent lie on a straight line parallel to the axis of the curve and passing through the point of contact of the tangent.

12. Find the equation to the chord through two points on an ellipse whose eccentric angles are given.

If P be a point on an ellipse, A, A' the extremities of the major axis, show that the tangent at P intersects the diameter parallel to AP on the tangent at A'.

13. **Prove geometrically that the product of the perpendiculars from the foci of an ellipse upon any tangent is equal to the square of the semi-axis minor.**

If one focus of an ellipse inscribed in a triangle be at the circumcentre, prove that the other is at the orthocentre of the triangle.

14. **Show that tangents from an external point to a conic subtend equal angles at a focus.**

If the tangent at a point *P* of a hyperbola meet an asymptote in *Q*, and *SP*, drawn from the focus *S* which lies within the branch of the curve on which *P* lies, meet the same asymptote in *R*, show that the triangle *RQS* is isosceles.

VII. MECHANICS.

[*Full marks may be obtained for about* two-thirds *of this paper. Great importance is attached to accuracy.* N.B.—*g may be taken* = 32 *feet per second per second.*]

1. One end, *B*, of a light cord is fixed; the cord passes over a fixed peg, *A*, in the horizontal line through *B*, and the other end, *C*, of the cord hangs down vertically below *A*. If a mass of weight *P* is suspended from *C*, find the magnitude and direction of the pressure on the peg.

2. Give the definitions of the *coefficient of friction* and the *angle of friction* between two bodies. Describe also any method by which these magnitudes have been measured.

A mass of 190 lbs. is placed on a rough inclined plane the tangent of whose inclination to the horizon is $\frac{5}{12}$, the coefficient of friction being $\frac{1}{3}$; find the magnitude of the horizontal force which will just suffice to drag the body up.

What is the magnitude of the horizontal force which will just prevent the body from sliding down?

3. Define the moment of a force about an axis (or a *point*, explaining when this latter expression may be used).

A and *B* are two fixed points in a given plane, 10 inches apart; a force *P* acts through *A*, but has any direction whatever in the plane; if *P* has always a moment of 60 inch-pounds' weight about *B*, give a graphic representation of the various magnitudes and directions of *P*.

What is the least, and what the greatest, value of *P*?

4. A ladder, *AB*, 15 feet long, rests against the ground at *A* and against a rough vertical wall at *B*, the coefficients of friction at *A* and *B* being $\frac{2}{3}$ and $\frac{1}{3}$ respectively; the centre of gravity, *G*, is 6 feet from *A*; find the inclination to the horizon at which the ladder will be just about to slip.

If the ladder is placed at an inclination $\tan^{-1}\frac{1}{2}$, and a boy whose weight is $\frac{1}{6}$ of that of the ladder ascends it in this position, how far will he be able to go before the ladder begins to slip?

5. Which of the accelerations, 15 miles per hour per 2 minutes and 2 inches per second per second, is the greater?

If a point moves from rest with the first of these, through how many feet will it move in 10 seconds?

6. If a particle slides down a rough inclined plane of inclination i, the coefficient of friction being μ, find its acceleration.

If $\tan i = \frac{3}{4}$, and $\mu = \frac{1}{2}$, in what time will the particle move from rest over 125 feet of the plane?

7. A mass of 8 ounces hangs from a spring balance in a balloon: what tension will be indicated by the balance,

(*a*) if the balloon is moving upwards with constant velocity;

(*b*) if it is moving upwards with an acceleration of 2 feet per second per second?

8. Enunciate the two principles on which the solution of the problem of the collision of two spheres depends.

Two spheres are approaching one another, their centres moving in one and the same straight line; their masses are 10 and 6 ounces, their respective velocities 40 and 60 feet per second, and their coefficient of restitution is $\frac{1}{5}$; find their velocities after collision.

If they are in contact for $\frac{1}{64}$ of a second, find, *in ounces' weight*, the magnitude of the mean pressure between them.

9. If a particle moves in a circle of radius r with a velocity v, prove that it has an inward normal acceleration equal to $\frac{v^2}{r}$.

If the mass of the particle is 2 ounces, the radius of the circle 8 inches, and the velocity at all points 16 feet per second, what is the magnitude of the resultant force acting on the particle in each position, and what is its precise direction?

Point out the erroneous conception involved in the term "centrifugal force."

10. State, in a general way, the kind of effect produced on the trajectory of a projectile by the resistance of the air.

If the resistance of the air can be neglected when a projectile is fired at an elevation $\tan^{-1}\frac{12}{5}$ with a velocity of 520 feet per second, when and where will the projectile strike an inclined plane passing through the point of projection, the inclination of this plane being $\tan^{-1}\frac{8}{15}$?

MATHEMATICAL EXAMINATION PAPERS

FOR ADMISSION INTO

Royal Military Academy, Woolwich,

JUNE, 1896.

OBLIGATORY EXAMINATION.

I. EUCLID.

[*Ordinary abbreviations may be employed; but the method of proof must be geometrical. Proofs other than Euclid's must not violate Euclid's sequence of propositions. Great importance will be attached to accuracy.*]

1. ABC and DEF are two triangles. If the sides BC, CA, AB are equal to the sides EF, FD, DE respectively, prove that the angle BAC is equal to the angle EDF.

2. ABC is any triangle, D any point inside the triangle. Prove that BD and DC are *together* less than BA and AC, and that the angle BDC is greater than the angle BAC.

The angle BAC is bisected by a straight line meeting BC in E, and P is any point on this straight line within the triangle. Prove that

(i.) BA and AC are *respectively* greater than BE and EC;

(ii.) BA and AC are *respectively* greater than BP and PC.

3. Write down Euclid's axiom with regard to *parallel straight lines*.

OA, OB are two finite straight lines meeting at O. C and D are any two points in OA and OB respectively. Through C and D straight lines are drawn at right angles to OC and OD respectively. Prove that these straight lines, if produced indefinitely in both directions, must meet.

4. *C* is a point in a finite straight line *AB*. Prove that the square on *AB* is equal to the squares on *AC* and *CB*, together with twice the rectangle contained by *AC* and *CB*.

Express this result algebraically.

5. *ABC* is an acute-angled triangle. *AD* is drawn perpendicular to *BC*. Prove that the squares on *AB* and *BC* are together equal to the square on *AC* and twice the rectangle contained by *BC* and *BD*.

If *M* is the middle point of *BC*, prove that the difference of the squares on *AB* and *AC* is equal to twice the rectangle contained by *BC* and *MD*.

6. If a chord and a diameter of a circle intersect at right angles, prove that the diameter bisects the chord.

AB is a diameter of a circle. *PQ* is a chord *not* at right angles to *AB*. *AM* and *BN* are drawn perpendicular to *PQ*. Prove that $PM = NQ$.

7. If two circles touch one another externally, prove that the straight line which joins their centres passes through the point of contact.

8. Show how to draw tangents to a circle from a given point outside it.

P and *Q* are points on a diameter of a circle produced, and are at equal distances from the centre. *PR* and *QT* are tangents from *P* and *Q*, the points *R* and *T* lying on opposite sides of *PQ*. Prove that *PRQT* is a parallelogram.

9. Prove that angles in the same segment of a circle are equal.

AB and *CD* are two intersecting chords of a circle. If the arcs *AD* and *BC* are together equal to the arcs *DB* and *CA*, prove that *AB* and *CD* are at right angles to one another.

10. Show how to describe a triangle whose sides shall touch a given circle and whose angles shall be equal to the angles of a given triangle.

11. Show how to find a mean proportional between two given straight lines.

12. *AB* is a diameter of a circle, *CD* is a chord at right angles to it. If any chord *AP* drawn from *A* cuts *CD* in *Q*, prove that the rectangle contained by *AP* and *AQ* is constant for different positions of *P*.

II. ARITHMETIC.

1. Simplify

$$\frac{17\frac{6}{7}}{9\frac{1}{11}}+2\tfrac{3}{7}\times\left(9\tfrac{1}{3}-2\tfrac{11}{17}\right)-\frac{5}{1-\frac{3}{31}}.$$

2. Divide ·1154255 by ·00115.

3. In how many years will £1500 amount to £1781. 5*s*. 0*d*. at 2½ per cent. per annum simple interest?

4. Find the value of ·30875 of a mile in furlongs, poles, and yards.

5. Express the sum of 7½ guineas 3½ crowns and 5¼ florins as a fraction of £53.

6. What is the rental of an estate of 645 acres 3 roods 25 poles at £2. 11*s*. 4*d*. per acre?

7. Find the least common multiple of 555, 1221, and 2035.

8. Find the cost of a carpet, for a room 24 feet long and 17½ feet wide, at 3*s*. 3*d*. per square yard, if a margin 2 feet wide be left uncovered.

9. A watch is offered for sale for £5. 15*s*. 0*d*.; and, if that price is reduced by 5 per cent., the dealer who is selling it will still make 9¼ per cent. profit: how much did the watch cost him?

10. State the rule for determining the remainder in a division sum when it is worked by dividing by two factors of the divisor successively. instead of by their product.

As an example, reduce 107 lbs. to quarters by dividing by 4 and by 7; and explain why you do what you do to find how many pounds there are over.

11. A man goes bankrupt with £1160 assets. His liabilities are the present value of three loans, at simple interest, amounting in all to £1800; one of them obtained 8 years ago at 4 per cent.; another, double of the first, 4 years ago at 5 per cent.; and the third, equal to the sum of the other two, a year ago at 8 per cent. How much does each of the three creditors lose?

12. A cistern 12 feet deep, of which the length is double the width, holds 21 tons of water: find the depth of another cistern which is $1\frac{3}{4}$ ft. shorter and 4 ft. wider than the other, and which holds the same quantity. One cub. ft. weighs 1000 ozs.

13. Taking the values of zinc and copper to be £17. 14*s.* 0*d.* and £73. 15*s.* 0*d.* per ton respectively, find how much of each metal there will be in a mass compounded of the two, which weighs 14 cwt., and is worth £23. 12*s.* 0*d.*

14. A ship, steaming towards a port in a fog, fires a signal gun which is answered from the port, as soon as heard, by another gun; the report of the latter reaches the ship $25\frac{1}{2}$ seconds after the first gun was fired; the ship fires again immediately, is answered as before, and this time the reply is heard in $24\frac{1}{2}$ seconds. If the sound travels 1100 feet per second, at what rate must the ship be steaming?

III. ALGEBRA.

(Up to and including the Binomial Theorem, and the theory and use of Logarithms.)

[N.B.—*Great importance will be attached to accuracy.*]

1. Remove the brackets in

$$7a+6[b-5\{c+4(b-3(a+2c))\}],$$

and find its value when $a=2$, $b=3$, $c=1$.

2. Prove that $a \times b = b \times a$, where a and b may have any positive integral values.

Multiply together

$$x-y, \quad x+y, \quad x^2-xy+y^2, \quad x^2+xy+y^2.$$

3. Show that $x^n - a^n$ is divisible by $x-a$ for all positive integral values of n.

Find the factors of

$$bc(b^2-c^2)+ca(c^2-a^2)+ab(a^2-b^2),$$

and

$$x^4+5x^3+5x^2-5x-6.$$

4. Prove that

$$x^4-15x^3+75x^2-145x+84 \text{ and } x^4-17x^3+101x^2-247x+210$$

have the same H.C.F. and the same L.C.M. as

$$x^4-13x^3+53x^2-83x+42 \text{ and } x^4-19x^3+131x^2-389x+420.$$

5. In an examination a candidate takes five compulsory and two optional papers, makes the same marks on each paper, and gets fifty marks too few to pass. On a second attempt he increases his marks on each paper in the ratio of 63 to 50, and omits one of the optional papers, thus securing a hundred marks more than before and passing. If all the papers have the same maximum, find what number of marks is required to pass.

6. Solve the equations

$$\frac{3x-16}{x-3}+\frac{2x+3}{x-2}=\frac{5x-2}{x+1},$$

$$3x^4-16x^3+26x^2-16x+3=0,$$

and form the quadratic equation, the sum of whose roots is 3, and the sum of the cubes of whose roots is 7.

7. If $\frac{a}{b}=\frac{c}{d}$ prove that a, b, c, d are proportionals according to Euclid's definition.

Find what number must be subtracted from each of the numbers, 8, 10, 13, 17, that the remainders may be in proportion.

8. Of any number of terms in Arithmetical Progression, show that the sum of the rth term from the beginning and the rth term from the end is equal to the sum of the first and last terms.

If $\frac{a}{b+c}$, $\frac{b}{c+a}$, $\frac{c}{a+b}$ are in A.P., show that a^2, b^2, c^2 are also in A.P.; a, b, c being positive quantities.

9. Find the total number of combinations which may be formed out of n things.

A pack of cards consisting of four suits of eight cards each is dealt to three players, so that each has ten cards and the remaining cards lie on the table. The eight cards of each suit being numbered in order from 1 to 8, find in how many ways the cards can be dealt so that one player may hold ten of the highest twelve cards, and another ten of the lowest twelve.

10. Prove that the coefficient of x^p in the expansion of $(1-x)^{-(n+1)}$ by the Binomial Theorem is equal to the coefficient of x^n in the expansion of $(1-x)^{-(p+1)}$ both expansions being in ascending powers of x.

Hence (or otherwise) find the sum of the first twelve coefficients in the expansion of $(1-x)^{-5}$ in ascending powers of x by the Binomial Theorem.

11. What is meant by the *characteristic* of a logarithm? Show that it may be determined by inspection in the case of logarithms to the base 10.

Find the number of integers in the product of 2^{19} by 3^{17}, and find the first four integers of the product.

12. At an election for eight representatives, each voter had seven votes. The candidates were six liberals, six conservatives, and two acceptable to both parties. All six liberals were defeated, and the average of their votes was 1596 less than the average of the votes of the six conservatives. Each of the non-party candidates obtained a number of votes equal to the sum of these two averages. Altogether 24,064 persons voted, there was no spoilt vote, and each voter gave his votes to seven different candidates. Assuming that the same proportion of each party voted for *both* the non-party candidates, but that otherwise the voting was on strict party lines, find (i.) what proportion of voters voted for both the non-party men, (ii.) how many of the voters were of each party, (iii.) the number of votes polled by the non-party candidates.

IV. PLANE TRIGONOMETRY AND MENSURATION.

[N.B.—*Great importance will be attached to accuracy.*]

1. Prove that the circumferences of circles vary as the radii.

Assuming the ratio of the circumference to the radius to be $6\frac{2}{7}$, find the circular measure of (*i.e.* number of radians in) one of the angles of a regular figure of 44 sides.

2. Trace the changes in the cosine of an angle as the angle varies from $0°$ to $360°$.

Show that the value of $\cos(A+B)$ found from the usual formula is of the right sign; where A and B are each obtuse and less than $135°$.

3. Find an expression for all angles which have a given cosine.

Find the three values of x, when $x^3-3x=2\cos 3A$.

4. Show that, if A is an acute angle,

$$2\sin\frac{A}{2}=\sqrt{1+\sin A}-\sqrt{1-\sin A};$$

and, assuming the value of $\sin 30°$, test by means of this formula and the Table of common logarithms the correctness of the value of $L\sin 15°$ as given in the Table of logarithmic sines.

5. Establish the identities:

(i.) $\cos^2 A+\cos^2 B+\cos^2(A+B)-2\cos A\cos B\cos(A+B)=1$;

(ii.) $\tan 50°-\tan 40°=2\tan 10°$;

(iii.) $\tan^{-1}\frac{1}{2}+\tan^{-1}\frac{1}{13}=\tan^{-1}\frac{3}{5}$.

6. Prove that, in general, small changes of the tabular logarithmic sine of an angle are approximately proportional to the corresponding changes of the angle.

7. In a plane triangle, prove that

(i.) $\tan\frac{B-C}{2}=\frac{b-c}{b+c}\cot\frac{A}{2}$;

(ii.) $\frac{\cos(B-C)}{\sin B\sin C}+\frac{\cos(C-A)}{\sin C\sin A}+\frac{\cos(A-B)}{\sin A\sin B}=4.$

8. If, in a plane triangle, $a = 447001$, $c = 341387$, $C = 37° 22' 12''$, find the two values of B; and draw a figure showing the two triangles obtained.

9. A balloon is vertically over a point which lies in a direct line between two observers who are 2000 feet apart, and who note the angles of elevation of the balloon to be $35° 30'$ and $61° 20'$; find its height.

10. (i.) Express the area of a regular polygon in terms of the radius of the circumscribed circle.

(ii.) Show that the cube of the perimeter of a triangle, multiplied by $\tan\frac{A}{2}\tan\frac{B}{2}\tan\frac{C}{2}$, is equal to the product of the diameters of the three escribed circles.

11. (i.) The *internal* diameter of a hollow ball of uniform thickness is two inches, and the thickness is one-fifth of an inch; find the number of cubic inches of material in the ball. [$\pi = \frac{22}{7}$.]

(ii.) Draw a plan and find, in acres, the area of a field, from the following notes:

	Yards	
	to E	
	550	
to C 55	242	154 to D
to B 176	110	
	from A	

V. STATICS AND DYNAMICS.

[*g may be taken* = 32 *feet per second per second.*]

1. Define force, and explain what are meant by the force of gravity and the tension of a string.

A weight of 10 lbs. hangs at the end of a uniform rope 12 feet long, whose other end is fixed. If the rope weighs 3 lbs., find its tension at the point 4 feet from the weight.

2. Enunciate the parallelogram of forces.

ABC is an equilateral triangle and D is the middle point of AB; prove that the resultant of forces represented in magnitude by AD, AC, is represented by $\sqrt{7}$ times AD.

3. Find the direction and magnitude of the resultant of two parallel forces acting in the same direction.

If a heavy body is partly supported by a string and partly by a smooth horizontal plane, prove that the string must be vertical.

4. Find the position of the centre of gravity of the area of a triangle.

If D is the middle point of one side BC of a triangle ABC, prove that the distance between the centres of gravity of the triangles ABD, ACD is one-third of BC.

5. Find what force, acting horizontally, will support a weight W resting on a smooth inclined plane, the base of the plane being three times its height; also find the pressure on the plane.

6. Find the ratio of the power to the weight in the system of pulleys in which the string round any pulley has one end fastened to a horizontal bar, from which the weight is suspended, and the other end to the next pulley, neglecting the weights of the pulleys and the bar.

If there are two movable pulleys, find the point on the bar from which the weight should be suspended.

7. Explain how uniform velocity and uniform acceleration are measured. Find the measure of the acceleration due to gravity when a yard and three seconds are the units of length and time.

8. Enunciate and prove the parallelogram of velocities.

A ship is sailing due north at the rate of three miles an hour, and a passenger on board walks transversely across the deck at the rate of 4·4 feet per second; find his actual directions of motion as he walks one way or the other.

9. A stone is projected vertically upwards with a velocity of 64 feet per second; neglecting the resistance of the air, find the greatest height to which it rises, and the time of its ascent and descent. Also find the times at which its height will be 28 feet.

10. State Newton's Second Law of Motion, and show how it gives a method of measuring force.

A body whose mass is 4 lbs. is moving with a velocity of 64 feet per second. If it is brought to rest in 128 feet by applying a constant resistance, find the magnitude of the resistance.

11. If a particle move with uniform velocity v in a circle of radius r, prove that its acceleration is in the direction of the centre of the circle, and is equal to $\frac{v^2}{r}$.

A heavy particle on a smooth horizontal plane is attached by a string to a fixed point on the plane. If the string be straightened, and the particle be projected horizontally in a direction perpendicular to the string, compare the tension of the string with the weight of the particle.

FURTHER EXAMINATION.

VI. PURE MATHEMATICS.

[*Full marks may be obtained for about* two-thirds *of this paper.*]

1. Find the length of a man's stride, if dividing the number of strides he makes per minute by 30 gives his speed in miles per hour.

Contrast the rate of striding of a runner who covers 100 yards in 9·8 seconds, with a stride of 7 feet, with the rate of pedalling of a bicyclist who, on a machine geared so as to be equivalent to one with a driving wheel of 80 inches diameter, rides 30 miles in one hour.

2. Indicate the method of solution of simultaneous linear equations.

Three trains, of lengths a, b, c (feet), are travelling with uniform velocities u, v, w (feet per second), in the same direction on equidistant parallel rails with their rear-most carriages in a straight line.

Show that the trains may all be cut by some straight line or other for a time

$$\frac{a+c}{2v-u-w} \text{ seconds,}$$

or, for a time,

$$\frac{2b}{u+w-2v} \text{ seconds,}$$

according as $2v \gtrless u+w$.

3. Prove that the number of permutations of n different things taken r at a time when each of the n things may be repeated is n^r.

In the decimal system of notation, how many numbers are there which consist of four digits? Prove that the sum of all such numbers is

49495500.

4. O is the centre of a circle, and AOB a diameter. Circles are described upon AO and OB as diameters. Show that the circle described, so as to touch the large circle internally and the two smaller circles externally, has a diameter one-third that of the larger circle.

5. $ABCD$ is a square of which BD is a diagonal. Through A a straight line is drawn, meeting BC and BD, produced, if necessary, in H and K.

If p be the perpendicular distance of K from BC, show that the reciprocal of p is equal to the sum or the difference of the reciprocals of BH and BA, and distinguish the cases.

If the line through A be drawn at random, show that it is an even chance that p is half the harmonic mean of BH and BA.

6. Construct a quadratic equation, with rational coefficients, so that one root may be

$$2 \sin 18^\circ.$$

7. Indicate the operations necessary in order to determine the distance between two inaccessible points in the same plane as the positions where the necessary observations are made.

8. Find the polar equation of a straight line in the form

$$r = p \sec(\theta - \alpha).$$

Find the condition that this straight line may touch the circle

$$r^2 - 2lr\cos(\theta - \beta) + l^2 - a^2 = 0.$$

9. Find the general equation of a circle whose centre lies on the axis of x.

If the abscissae of the centres of two such non-intersecting circles be $+a$ and $-a'$, and their radii r and r', find the coordinates of points on the axis of x at which the circles subtend equal angles. Find the equation of the radical axis of the two circles.

10. Taking the principal axes of an ellipse as coordinate axes find its equation.

If the ellipse be rotated through an angle θ in the positive direction show that its equation becomes

$$\cos^2\theta\left(\frac{x^2}{a^2}+\frac{y^2}{b^2}\right)+\sin^2\theta\left(\frac{x^2}{b^2}+\frac{y^2}{a^2}\right)+\sin 2\theta \, . \, xy\left(\frac{1}{a^2}-\frac{1}{b^2}\right)=1.$$

If the direction of rotation be reversed, how is this equation affected?

11. Find the equation of the normal to an ellipse at a given point.

The equation of an ellipse is $\frac{x^2}{4}+y^2=1$. Find the coordinates of the intersection of normals at the points whose eccentric angles are 75° and 15°.

12. Find the equation of a hyperbola in rectangular coordinates.

Show that if a variable line form, with two fixed lines, a triangle of constant area, the locus of a point which divides the intercept made on the variable line in a given ratio, is a hyperbola.

13. In a parabola, prove that an isosceles triangle is formed by the focal distance of a point, the normal at the point and the axis.

Find the locus of the foot of the perpendicular from the focus on the normal.

(*This question is to be solved geometrically.*)

14. Prove that the feet of the perpendiculars from the foci on any tangent to an ellipse lie on a circle whose radius is equal to the semi-major axis.

In an ellipse, if a line be drawn through a focus making a constant angle with the tangent, prove that the locus of the point of intersection with the tangent is a circle. Find also its centre and radius.

(*This question is to be solved geometrically.*)

VII. MECHANICS.

[*Full marks may be obtained for about* two-thirds *of this paper. Great importance is attached to accuracy.* N.B.—*g may be taken* = 32 *feet per second per second.*]

1. Explain the derivation of the triangle of forces from the parallelogram of forces.

Two forces are represented by the lines joining the middle points of opposite sides of a quadrilateral. Show that their resultant is represented in magnitude and direction by one of its diagonals.

2. Show that if any number of coplanar forces acting at a point are in equilibrium, the sum of their components resolved in any two directions at right angles to each other must each be zero.

A string 31 inches long passes through a small ring of 4 ounces' weight, and has its extremities fixed at two points 25 inches apart, and in the same horizontal line. Find the tension of the string in the position of equilibrium in ounces' weight correct to two places of decimals.

Find also the magnitude of the horizontal force which, applied to the ring, will cause it to rest at a point 7 inches from the nearer end of the string.

(*This question may be solved graphically or analytically.*)

3. Prove that a system of coplanar forces will be in equilibrium if the algebraical sum of the moments of the forces about any three points not in the same straight line vanishes in each case.

A triangular lamina ABC, whose sides BC, CA, AB are respectively 18, 24, and 30 inches in length, is placed in a vertical plane with BC, CA resting upon two fixed smooth pegs D, E, 20 inches apart and in the same horizontal line. If equal weights W, W be suspended from A and B, and the triangle be kept with AB horizontal by means of a string connecting C with the peg D; find the tension of the string and the pressures on the pegs, neglecting the weight of the lamina.

(*This question may be solved graphically or analytically.*)

4. Define the centre of mass (centre of gravity) of a heavy body, and show that if a heavy body be suspended from a fixed point its centre of mass must be vertically beneath the point.

In a lamina of any form a line AB of length c is taken, and it is observed that when the lamina is suspended from A, the line AB dips 30° below the horizon, and 45° when suspended from B. Find the distance of the centre of mass of the lamina from AB.

(*This question may be solved graphically or analytically.*)

5. A uniform ladder rests with one end against a vertical wall and the other on the ground, inclined to the vertical at 45°. Compare the least horizontal forces which, applied to the foot of the ladder, will move it towards or from the foot of the wall, the coefficient of friction being the same for both ends of the ladder.

6. A particle projected with a velocity u moves subject to an acceleration α in the direction of motion. Find the space described in the nth unit of time.

A particle slides from rest down a smooth plane inclined at 30° to the horizon. Find the position of that length of 92 feet which is passed over by the particle in one second.

7. Two scale pans, each of 4 oz. mass, are connected by a light inextensible string, which passes over a smooth fixed pulley. If a mass of 2 oz. be placed in one pan, and a mass of 3 oz. in the other, find the tension of the string and the pressures of the masses on the scale pans.

8. Two spheres of elasticity e and masses m, m', moving with velocities u, u', impinge directly. Find their velocities after impact.

The centres of two equal billiard balls of radius a and elasticity $\frac{5}{8}$ move along the straight lines, whose equations are in rectangular coordinates

$$4y = 3(x - a),$$
$$5y = -12(x + a),$$

in such a manner that the line joining them is always parallel to the axis of x, and impinge at the origin. Find the equations of their lines of motion after impact.

9. Prove that the path of a projectile in vacuo is a parabola, and that the velocity at any point is that due to falling from the directrix.

10. A particle of mass m attached to a fixed point by a light string of length l, makes complete revolutions in a vertical plane under the action of gravity. If u be its velocity at the highest point, find its velocity in any other position, and also the tension of the string.

If the ratio of its maximum to its minimum velocity be $a : b$, show that the maximum tension of the string will be to the minimum in the ratio of $5a^2 - b^2$ to $5b^2 - a^2$.

MATHEMATICAL EXAMINATION PAPERS

FOR ADMISSION INTO

Royal Military Academy, Woolwich,

NOVEMBER, 1896.

OBLIGATORY EXAMINATION.

I. EUCLID.

[*Ordinary abbreviations may be employed, but the method of proof must be geometrical. Proofs other than Euclid's must not violate Euclid's sequence of propositions. In the absence of special directions to Candidates, any of the propositions within the limits prescribed for examination may be used in the solution of problems and riders. Great importance will be attached to accuracy.*]

1. Give accurate definitions of the following geometrical terms:—superficies, rectilineal angle, circle, rhombus, postulate, axiom.

2. Draw a straight line perpendicular to a given straight line of unlimited length, from a given point without it.

Construct a square which shall have an extremity of one of its diagonals at a given point, and the extremities of the other diagonal on a given straight line.

3. Prove that if a side of any triangle be produced, the exterior angle is equal to the two interior and opposite angles, and that the three interior angles of every triangle are together equal to two right angles.

From a vertex of an equilateral triangle a perpendicular is drawn to the opposite side, and upon this perpendicular another equilateral triangle is constructed; show that its sides are perpendicular to those of the original triangle.

4. Prove that in obtuse-angled triangles, if a perpendicular be drawn from either of the acute angles to the opposite side produced, the square on the side subtending the obtuse angle is greater than the squares on the sides containing the obtuse angle, by twice the rectangle contained by the side on which, when produced, the perpendicular falls, and the straight line intercepted without the triangle, between the perpendicular and the obtuse angle.

Prove that a triangle, the sides of which are three, four, and six inches in length, is obtuse-angled.

5. Construct a square which shall be equal to a given rectilineal figure.

6. Prove that the diameter is the greatest straight line in a circle, and that, of all others, that which is nearer to the centre is always greater than one more remote.

7. Prove that the opposite angles of any quadrilateral figure inscribed in a circle are together equal to two right angles.

8. On a given straight line describe a segment of a circle, containing an angle equal to a given rectilineal angle.

Through a given point draw a straight line which shall cut off from a given circle a segment containing an angle equal to a given rectilineal angle.

9. Construct an isosceles triangle, having each of the angles at the base double of the third angle.

Prove that the perpendicular drawn to one side of such a triangle, at its middle point, cuts the other side in extreme and mean ratio.

10. Prove that the sides about the equal angles of triangles which are equiangular to one another are proportionals.

The altitude of a certain triangle is equal to its base; show that if a rectangle be inscribed in it so as to have one side along the base and the extremities of the opposite side upon the sides of the triangle, then the three triangles by which the original triangle exceeds the rectangle are together equal to half the square on the diagonal of the rectangle.

11. Prove that parallelograms which are equiangular to one another have to one another the ratio which is compounded of the ratios of their sides.

12. Prove that, in any right-angled triangle, any rectilineal figure described on the side subtending the right angle is equal to the similar and similarly described figures on the sides containing the right angle.

II. ARITHMETIC.

1. Simplify

$$\frac{7\frac{1}{3}+3\frac{1}{7}}{11\frac{1}{9}-9\frac{1}{11}}\times\left(2\frac{3}{4}-\frac{3}{8\frac{1}{2}+6\frac{1}{6}}\right).$$

2. Divide 26·751 by ·000925.

3. Find the Least Common Multiple of 34, 42, 119, and 255; and obtain the sum of the fractions $\frac{3}{34}$, $\frac{1}{42}$, $\frac{10}{119}$, and $\frac{1}{255}$, by reducing them to a common denominator.

4. Find the value of $\frac{4}{15}$ of £1. 19*s*. 8¼*d*.

5. Find the cost of 25 quarters 3 bushels of oats at 19*s*. 3½*d*. a quarter.

6. One side of a rectangular field is $\cdot 0\dot{5}4$ of a mile, and the adjacent side is $\cdot 1\dot{3}$ of a furlong. Find the length of each side in yards and feet, and express the area of the field as a fraction of an acre.

7. A sum of £377. 13*s*. 4*d*. lent at simple interest amounts in a year and a half to £396. 11*s*. 0*d*. What is the rate of interest?

8. A garden whose length is 67 ft. 9 in. has a path 4 ft. wide on the two sides and at one end: if it costs £4. 10*s*. 3¾*d*. to turf the remainder at 6*d*. a square yard, what is the width of the garden?

9. *A*, *B*, and *C* who are engaged on piece-work do amounts in the same time which bear to one another the proportion of 10, 9, and 14 respectively. What ought each to receive if the amount paid for the whole is £24. 15*s*.?

10. What is meant by an *odd* number and an *even* number?

Show that it is only necessary to look at the digit in the unit's place to ascertain if a number is odd or even.

Several numbers have to be added together. What is the condition that their sum should be odd?

11. A cistern is filled in 3½ hours by a pipe 3 sq. in. in cross section through which water flows at the rate of 6·4 miles an hour. What is the volume of the cistern?

12. If coffee and chicory cost £8. 10*s.* and £2. 10*s.* per cwt. respectively, what is the proportion of coffee and chicory in a mixture of which 7 lbs. are worth 7*s.* 6*d.*?

13. A man buys goods and finds that the cost of carriage is 4 per cent. on the cost of the goods. He is compelled to sell at a loss of 5 per cent. on his total outlay; if however he had received £3. 5*s.* more than he did he would have gained $2\frac{1}{2}$ per cent. What was the original cost of the goods?

14. *A* and *B* set out from the same place in the same direction and travel uniformly; after 9 days' travelling *A* finds he is 72 miles ahead of *B*: he then turns and travels back the distance *B* would travel in 9 days, he then turns again and overtakes *B* in $22\frac{1}{2}$ days from the start. What is the rate of travelling of each?

(Up to and including the Binomial Theorem, and the theory and use of Logarithms.)

[N.B.—*Great importance will be attached to accuracy.*]

1. What are the factors of x^6+y^3?

Hence, or otherwise, show that

$$(1+6a+6a^2)^6+(3+12a+12a^2)^3 \text{ is divisible by } (2+6a+6a^2)^2;$$

and find the value of the quotient when $a = -\frac{1}{2}$.

2. Find three factors of

$$1-x^2-2x^3-2x^4-x^5+x^7;$$

and the H.C.F. of

$$6x^3-13x^2+19x-7 \text{ and } 9x^3-27x^2+41x-28.$$

3. Simplify

$$\frac{1}{y\left(\frac{x}{y}-\frac{y}{x}\right)}-\frac{1}{x\left(\frac{x}{y}+\frac{y}{x}\right)}+\frac{1}{\left(\frac{1}{x}+\frac{1}{y}\right)(x^2+y^2)};$$

and prove that $\sqrt{9\sqrt{6}+6\sqrt{12}}+\sqrt{9\sqrt{6}-6\sqrt{12}}$

is equal to $2\sqrt[4]{216}$.

4. Solve the equations:

(i.) $6x = 25(x^{\frac{1}{2}}-1)$;

(ii.) $\begin{cases} 4x^2-6xy+27y^2 = 5, \\ 8x^2-9y^2 = 1; \end{cases}$

and show that, if x be real, $\dfrac{x^2-50x+625}{x-50}$ can have no positive value less than 100.

5. I sent cash to a grocer for a certain number of lbs. of sugar, at the rate of 7 lbs. for 1*s.* $1\frac{1}{2}$*d.* But before the order reached him the price of sugar had risen, and the money was sufficient only to buy a quantity less by $10\frac{1}{2}$ lbs. than that which I had intended; so I sent an additional 5*s.* $7\frac{1}{2}$*d.*, and received one-fifth as much again as I had at first ordered. Find the number of lbs. ordered at first, the rise in the price being less than a half-penny a pound.

6. Show that, in a scale of notation of which the radix is r, when the sum of the digits of any whole number is divided by $r-1$, there will be the same remainder as when the whole number is divided by $r-1$.

If A and B be numbers in the scale of 10, and if 5 and 7 be the remainders when the sums of the digits of A and B are respectively divided by 9; find the remainder when the sum of the digits of the product of A and B is divided by 9.

7. Find expressions for the sum and general term of an Arithmetical Progression.

In a series of right-angled triangles, one side has, in succession, the lengths 1×4 in., 2×6 in., 3×8 in., 4×10 in., etc., and the hypotenuse differs in length from this side by one inch; show that the lengths of the other side form an Arithmetical Progression.

8. Find the number of permutations of n dissimilar things taken r together.

Find the number of arrangements of three different letters which can be formed of the ten letters Q to Z; Q, when it occurs, being always followed by U.

9. Prove the Binomial Theorem when the index is a positive fraction, assuming its truth for a positive integer.

Calculate by logarithms the numerical value of the 6th term of the expansion, by the Binomial Theorem, of $(1-x)^{120}$, when $x = \cdot 220793$; and show that it is nearly 10^5.

10. Write down the expansions for a^x, e^x and $\log_e(1+x)$, each in a series of ascending powers of x.

If $z = \cdot 9999999999$, and $e = 2\cdot 71828$; find, to four places of decimals, the value of $z + \frac{1}{2}z^2 + \frac{1}{3}z^3 +$ etc.

IV. PLANE TRIGONOMETRY AND MENSURATION.

[N.B.—*Great importance will be attached to accuracy.*]

1. Show that the angle subtended by any arc of a circle at the centre, may fairly be measured by the ratio of the arc to the radius of the circle.

Why may it not be measured by the ratio of the chord of the arc to the radius of the circle?

2. Define the tangent of a positive angle less than 360°.

If the angle A be known to be positive and less than 360°, what possible values can it have when

(i.) $\tan A = \sqrt{3}$. (ii.) $\tan \frac{1}{2}A = -1$. (iii.) $\tan 2A = \tan A$?

3. Prove the expression for $\sin(A+B)$ in terms of $\sin A$, $\cos A$, $\sin B$, $\cos B$, showing that the result is true for all sizes of the angles A, B. Hence find all the trigonometrical ratios of 105°.

4. Prove that in any triangle the sides are proportional to the sines of the angles opposite to them; and that the cosine of any angle of the triangle is expressible, in terms of the sides, by the formula

$$2bc \cos A = b^2 + c^2 - a^2.$$

If the sides of a triangle be 4, 5, 6, find the cosines of the angles; and hence, with the table of logarithms, determine the smallest angle to the nearest minute.

5. In any triangle ABC prove that

$$\tan \tfrac{1}{2}(B-C) = \frac{b-c}{b+c} \cot \tfrac{1}{2}A.$$

If $b = 11$, $c = 3$, $A = 57°$, find B and C to the nearest second, using the table of logarithms.

6. Prove the relations

(i.) $\tan A + \tan B + \tan C = \tan A \tan B \tan C$ where $A + B + C = 180°$;

(ii.) $\sin^2 B \sin^2(C-A) + \sin^2 C \sin^2(A-B)$
$+ 2 \sin B \sin C \sin(C-A) \sin(A-B) = \sin^2 A \sin^2(B-C)$.

7. What is the meaning of $\tan^{-1}x$?

Show that

(i.) $\cos^{-1}[xy - \sqrt{1 - x^2 - y^2 + x^2y^2}] = \cos^{-1}x + \cos^{-1}y,$

(ii.) $\tan^{-1}\left(\dfrac{x+y+z-xyz}{1-yz-zx-xy}\right) = \tan^{-1}x + \tan^{-1}y + \tan^{-1}z.$

8. If D be the orthocentre of the triangle ABC, that is, the point of intersection of the perpendiculars drawn from A, B, C to the opposite sides, prove that the radius of the circle drawn through A, B, C is such that

$$R = \frac{BC}{2\sin A} = \frac{AD}{2\cos A}.$$

Prove also that, if O be the centre of the circle drawn through A, B, C,

$$OD^2 = R^2(1 - 8\cos A\cos B\cos C).$$

9. Prove that the area of any quadrilateral $ABCD$ is given by

$$\sqrt{(s-a)(s-b)(s-c)(s-d) - abcd\cos^2\tfrac{1}{2}(A+C)},$$

where $a = AB$, $b = BC$, $c = CD$, $d = DA$, $2s = a+b+c+d$, and A, C are the angles DAB, BCD respectively.

10. Prove that the area of a triangle is half the product of the base and height, and that the volume of a right circular cone is one third the product of the area of the base, and the height.

What is the height of a right circular cone when its volume is equal to that of a sphere of which the radius is equal to the radius of the base of the cone?

V. STATICS AND DYNAMICS.

[*g may be taken* = 32 *feet per second per second.*]

1. When is a force said to be given? Explain why a force can be represented by a finite straight line.

2. Assuming the parallelogram of forces, enunciate and prove the triangle of forces.

Two forces are represented in magnitude and direction by AO, BO; show that their resultant is represented by twice CO, where C is the middle point of AB.

If the forces are represented by AO, OB; find the resultant.

3. Two parallel forces whose magnitudes are P and Q act at the points A, B, of a body, in the same direction, and the length AB is given. Determine the point at which the resultant intersects the straight line AB.

The extremities A, B of a uniform rod rest on two supports while a weight of 30 pounds is fastened at a point C such that AC is twice CB. Find the pressure at A, (1) when the rod is light, (2) when it has a weight of ten pounds.

4. Find the centre of gravity of a uniform lamina bounded by a parallelogram. Show that it coincides with that of four particles of equal mass placed at the angular points.

5. A weight W is placed on a smooth inclined plane whose inclination to the horizon is two-thirds of a right angle, and is supported by a force whose line of action makes with the horizon an angle half that of the plane. Find the magnitude of the force.

6. Find the relation of the power to the weight in a single moveable pulley, (1) when the straight portions of the string are parallel, and (2) when they are at right angles.

7. Explain how velocity is measured, (1) when uniform, and (2) when variable. Compare the velocities of two trains, when one moves at the rate of 45 miles per hour, and the other at the rate of 33 feet per second.

8. A weight W is suspended from a point O by a light string. Find the tension when the point O ascends vertically, (1) with a uniform velocity, and (2) with a uniform acceleration equal to one-third that of gravity.

9. The diameter AB of a given circle is vertical, and A is the lowest point. Prove that the times of descent from rest, down all smooth chords ending at A are equal.

Prove also that the velocity of arrival at A is proportional to the length of the chord.

10. A body is projected vertically upwards with a velocity of 640 yards per minute; find the height at which the velocity is 16 feet per second.

The point A is vertically over B and the distance AB is 30 feet. Two particles are simultaneously projected, one downwards from A, and the other upwards from B, each with a velocity of 30 feet per second. Find the distance from the middle point of AB at which they meet.

11. Enunciate the laws of motion.

A particle moving in given direction AB, with a velocity of 32 feet per second, is acted on by a uniform transverse force whose acceleration is g and whose direction AC is perpendicular to AB. Find the position of th particle and its direction of motion at the end of one second.

FURTHER EXAMINATION.

VI. PURE MATHEMATICS.

[*Only* 10 *questions to be answered.*]

1. Show that the problem of finding the number of terms of an arithmetical progression when the first term, the common difference and the sum are given, may have more than one solution. If the sum of n_1 terms be equal to the sum of n_2 terms, show that the number n_1+n_2 depends only upon the ratio of the first term to the common difference.

2. Prove the Binomial Theorem for a positive integral index.

Express in its simplest form the coefficient of x^n in the expansion of the fraction

$$\frac{1}{(1+x+x^2+x^3+x^4)^3}$$

in ascending powers of x.

3. Investigate the expansion of $\log(1+x)$ to base e in a series of ascending powers of x.

Obtain a formula which would be convenient for the calculation of $\log_e 2$.

Explain how to pass from $\log_e 2$ to $\log_{10} 2$.

4. Divide a circle into two segments so that the angle contained in one shall be seven times the angle contained in the other, proving the proposition on which the construction depends.

5. Prove that the three perpendiculars from the angles of a triangle to the opposite sides intersect in one point. If H be the intersection and D, E, F the feet of the perpendiculars from the angles A, B, C, respectively, show that the four points A, B, C, H are the centres of the four circles that can be described so as to touch the sides of the triangle DEF.

6. Investigate an expression for the cosine of an angle of a triangle in terms of the sides.

If θ, ϕ be the greatest and least angles of a triangle, the sides of which are in Arithmetical Progression, show that

$$4(1-\cos\theta)(1-\cos\phi)=\cos\theta+\cos\phi.$$

7. In order to determine the height of a mountain a base was measured of 2750 feet. At either extremity of the base were taken the angles formed by the summit and the other extremity. These angles were 58° 28′ and 111° 53′. Also at the extremity from which the latter angle was taken the angular height of the mountain was 11° 19′.

Find the height of the mountain.

8. Show how to construct a regular pentagon. Choose the most convenient axes and write down the equations of its sides and of its inscribed circle.

9. Find the angle between the straight lines represented by

$$x^2+y^2=2xy\operatorname{cosec}\alpha,$$

the axes being rectangular.

Find the area of the triangle formed by these lines and the line

$$lx+my+n=0.$$

10. Find the equation to the tangent to the parabola

$$y^2-4ax=0.$$

Trace the parabolas

$$x^2=8ay,\quad y^2=27ax$$

and show that they intersect at right angles and also at an angle $\tan^{-1}\frac{9}{13}$.

11. From P, a point on an ellipse whose centre is C, an ordinate PM is drawn parallel to the minor axis. On PM a point Q is taken such that $QC=PM$.

Find the locus of Q.

12. Establish an equation to a hyperbola. Prove that the eccentricity is the secant of half the angle between the asymptotes.

13. In an ellipse S, F are the foci and P, Q any two points on the curve. If tangents at P, Q intersect at T, show that the angles QTS, PTF are equal.

14. Prove that the rectangle contained by the intercepts made by any tangent to a hyperbola on its asymptotes is constant.

VII. MECHANICS.

[*Full marks may be obtained for about* two-thirds *of this paper. Great importance is attached to accuracy.* N.B.—*g may be taken* = 32 *feet per second per second.*]

1. Three forces proportional to the sides of a triangle act at the middle points of these sides in directions perpendicular to them. Prove that the forces will be in equilibrium if they act all inwards or all outwards. Forces $k \, . \, AB$, $k \, . \, BC$, $k \, . \, CD$, and $k \, . \, AD$ act perpendicularly to the sides AB, BC, CD, DA of a quadrilateral at their middle points, the first three acting inwards and the last outwards. Find their resultant.

2. Explain how to find the resultant of a number of parallel forces applied at given points in a straight line.

A rectangular portmanteau 3 feet in length and 2 feet in height and weighing 56 lbs. is carried up a staircase by two men supporting it along the front and back edges of its bottom face. If this face be held at an inclination of 30° to the horizon, find, to two decimal places of a pound, what portions of the weight are supported by the two men, supposing the centre of mass of the portmanteau to be at its centre of figure.

3. A body is made up of a number of different portions whose centres of mass are all in one plane. If the positions of these centres of mass are given and also the weights of the several portions, find the position of the centre of mass of the whole body.

Prove that the centre of mass of any quadrilateral lamina may be found by the following construction:—Let E, F be the middle points of the diagonals of the quadrilateral, O their point of intersection. On OE, OF cut off $OM = \frac{2}{3} OE$ and $ON = \frac{2}{3} OF$ and complete the parallelogram $OMGN$ having OM, ON as adjacent sides. Then G will be the centre of gravity of the lamina.

(*This question may be solved either geometrically or analytically.*)

4. Find the mechanical advantage in the common screw-press working without friction.

The coupling between two railway carriages consists of a right-handed and left-handed screw at opposite ends of a long bolt, working in nuts attached to the two carriages, each screw having five threads to the inch.

Find the force which when applied to the end of an arm 18 inches long will cause the carriages to be drawn together with a force of 1 ton, taking the ratio of the circumference to the diameter of a circle as $3\frac{1}{7}$.

5. Explain what is meant by the *angle of friction.*

A body is placed on a rough plane whose inclination to the horizon is 45° the angle of friction being 30°. Prove that the force which must be applied to the body in a horizontal direction in order to just prevent it from slipping down is the same as if the plane were made smooth and its inclination to the horizon were decreased to 15°.

(*This question may be solved either geometrically or analytically.*)

6. A slip-carriage is detached from a train and brought to rest with uniform retardation in two minutes during which time it travels two-thirds of a mile. With what velocity was the train travelling when the carriage was detached?

The weight of the whole train was 130 tons and that of the slip-carriage 10 tons. Before the carriage was detached the train was just kept going with uniform velocity, by the pull of the engine, the resistance due to friction amounting to 35 lbs. weight per ton. Supposing the engine to pull the train with the same force after the carriage is detached, find the acceleration.

7. In Atwood's machine, two weights of 5 lbs. and 7 lbs. are attached to opposite ends of a string which passes over a light smooth pulley, and a rider weighing 4 lbs. is placed over the smaller weight. Find the acceleration.

When the system has moved through one foot from rest, the rider is detached by coming in contact with a fixed ring. How far will the 5 lb. weight descend below the ring before coming to rest?

8. A mass of m lbs. is making n complete revolutions per second on a smooth horizontal table in a circle of radius r to whose centre it is connected by a string. Find the tension of the string expressed in pounds' weight.

In the Watt's governor of a steam-engine, two equal light rods OA, OB, 1 foot long, are freely hinged at one end O to a vertical axis, and to their other ends A, B are attached equal weights of 7 lbs. The whole system is revolving about the vertical axis so that the rods OA, OB make equal angles with the vertical on opposite sides of it. Taking $\pi = \frac{22}{7}$, find the number of revolutions per minute when the rods include an angle of 120°.

9. A particle is projected *in vacuo* from a point O with a velocity whose horizontal and vertical components are 32 feet per second and 80 per second (upwards) respectively. Draw a careful figure exhibiting the positions of the particle after 1, 2, 3, 4, and 5 seconds respectively, and state the coordinates of the particle at these instants, referred to horizontal and vertical axes through O.

Find also the coordinates of the focus of the parabola described.

10. A sphere of mass M moving with velocity U overtakes and impinges directly on a sphere of mass m, moving with velocity u in the same straight line. If the elasticity of the spheres be perfect, find by how much the kinetic energy of the first sphere is decreased and that of the second increased.

Three equal spheres whose coefficient of restitution (*i.e.* elasticity) is $\frac{1}{2}$ are ranged at rest in a straight line. The first sphere is then projected in the same straight line with velocity U, so as to impinge directly on the second. Find the final velocities of the three spheres after the last collision has taken place between them.

MATHEMATICAL EXAMINATION PAPERS

FOR ADMISSION INTO

Royal Military Academy, Woolwich,

JUNE, 1897.

OBLIGATORY EXAMINATION.

I. EUCLID.

[*Ordinary abbreviations may be employed, but the method of proof must be geometrical. Proofs other than Euclid's must not violate Euclid's sequence of propositions. In the absence of special directions to Candidates, any of the propositions within the limits prescribed for examination may be used in the solution of problems and riders. Great importance will be attached to accuracy.*]

1. Define carefully the following geometrical terms:—point, plane, plane angle, circle, parallelogram.

2. Prove that if two angles of a triangle are equal to one another the sides which are opposite to the equal angles are also equal to one another.

3. Prove that if the square described on one of the sides of a triangle is equal to the sum of the squares described on the other two sides of it, the angle contained by these two sides is a right angle.

4. Prove that if a straight line be divided into any two parts, the squares on the whole line, and on one of the parts, are together equal to twice the rectangle contained by the whole and that part, together with the square on the other part.

5. Prove that in every triangle the square on the side subtending an acute angle is less than the sum of the squares on the sides containing that angle, by twice the rectangle contained by either of these sides, and the projection on it of the other side.

6. Prove that if two circles touch each other externally, the straight line which joins their centres passes through their point of contact.

7. Prove that if a straight line touch a circle the straight line drawn from the centre to the point of contact is perpendicular to the line touching the circle.

8. Prove that if from any point without a circle two straight lines be drawn, of which one cuts the circle and the other touches it; the rectangle contained by the whole line which cuts the circle, and the part of it without the circle, is equal to the square on the line which touches it.

9. Inscribe a circle in a given triangle.

10. Prove that in a right-angled triangle, if a perpendicular be drawn from the right angle to the base, the triangles on each side of it are similar to the whole triangle, and to one another.

11. Prove that similar triangles are to one another in the duplicate ratio of their homologous sides.

12. Prove that in equal circles angles at the centre have the same ratio which the arcs on which they stand have to one another.

13. If two sides of a triangle are produced and the exterior angles between the produced sides and the third side are equal to one another; prove, without using any proposition subsequent to Euclid I. 6 (question 2), that the triangle is isosceles.

14. Within a parallelogram $ABCD$ any point K is taken; a straight line drawn through K parallel to AB meets AD and BC at H and G respectively, and a straight line drawn through K parallel to AD meets AB and CD at E and F respectively. Prove that the squares on EH and FG together exceed or fall short of the sum of the squares on HF and EG by a rectangle independent of the position of K.

15. Construct a square of which one side shall lie along a given tangent to a given circle and of which the opposite side shall be a chord of the circle.

16. The circle inscribed in the triangle ABC touches the side BC at D. Show that the circles inscribed in the triangles BAD and CAD touch each other.

17. Through the extremities of two fixed perpendicular radii of a circle a pair of parallel chords are drawn. Prove that the circle whose diameter is the line joining the middle points of the two parallel chords passes through a fixed point, and also prove that the centre of this circle lies upon a fixed circle.

II. ARITHMETIC.

Arithmetical methods of solution are required, and all the working as well as the answers must be shown.

1. Find the cost of 49 cwt. 3 qrs. 14 lbs. at £13. 8*s*. 10*d*. per cwt.

2. Divide 5·4 by 0·00072 and the quotient by 1,470,568 to four places of decimals.

3. Find the greatest common measure of 27531 and 8740.

4. Add $\frac{1}{51}+\frac{2}{45}+\frac{3}{38}+\frac{4}{85}+\frac{5}{57}$.

5. Find (in its lowest terms) the difference between $\frac{1638}{2205}$ and $\frac{910}{1911}$.

6. The French metre is 3·281 English feet, the Rhenish foot is 12·356 English inches. Express 4·567 metres in terms of Rhenish feet to 3 places of decimals.

7. A sum of money put out at Simple Interest amounts to £688 when the rate is $2\frac{1}{2}$ per cent. and the time 3 years. What would the amount be if the rate were $3\frac{1}{4}$ per cent. and the time $2\frac{1}{8}$ years?

8. Show that if a square number ends in 6 the figure in the tens' place is odd; if it ends in any other number, the tens' figure is even.

9. If *B*'s wages are $\frac{7}{10}$ths of *A*'s, for how many days will a sum which pays *A*'s wages for 119 days pay the wages of both together?

10. If the manufacturer makes a profit of 25 per cent., the agent one of 8 per cent., and the shopkeeper one of 20 per cent., what is the cost to the manufacturer of an article which is sold in the shop for £32. 8s.

11. A man buys an article and sells it at a profit of 10 per cent. If he had bought it at 10 per cent. less, and sold it for 6*d*. less, he would have made a profit of 20 per cent. Find the cost price.

12. To do a piece of work, a contractor can employ two classes of workmen, whose wages are in the ratio of 17 to 13. If he employs the higher paid men (who work the faster) he pays £148. 15*s*. in wages, being £10. 10*s*. less than the lower paid men would cost him. Compare their rates of work.

13. A milkman adulterates his milk as follows:—He takes a pint out of each gallon and replaces it with water. He then takes another pint out of the mixture and replaces that with water; and he repeats the operation a third time. If he then sells his adulterated milk at the price per gallon that the pure milk originally cost him, what is his gain per cent.?

14. A tradesman makes up his receipts monthly. His receipts for the first two months average £3 less than those of the first month; those of the first three months average £4 more than the average of the first two; of the first four £5 more than that of the first three; of the first five £2 less than that of the first four; and of the first six £7 more than that of the first five. If the average monthly receipts for the first six months are £611, what were the receipts for each month?

III. ALGEBRA.

(Up to and including the Binomial Theorem, and the theory and use of Logarithms.)

[N.B.—*Great importance is attached to accuracy.*]

1. Show that

$$x^3y^2 + y^3z^2 + z^3x^2 - x^2y^3 - y^2z^3 - z^2x^3$$

is divisible by each of the factors $x-y$, $y-z$, $z-x$, and find the expression of the second degree which is the other factor.

Find also the factors of

$$1 - b - a^2 + a^3b + a^2b^3 - a^3b^3.$$

2. Prove that if

$$A = bB + C,$$
$$B = cC + D,$$
$$C = dD + E,$$
$$D = eE,$$

all symbols representing rational integral algebraic functions of x, then E is the highest common factor of A and B, considered as functions of x.

Find the highest common factor of

$$12x^3 - 16x^2 - 73x + 39 \quad \text{and} \quad 18x^3 + 51x^2 + 20x - 13,$$

and write down what the two expressions and their highest common factor become when $x = 10$.

3. Find the product of

$$\sqrt{2+\sqrt{3}}, \quad \sqrt{2+\sqrt{2+\sqrt{3}}}, \quad \sqrt{2+\sqrt{2+\sqrt{2+\sqrt{3}}}}, \quad \sqrt{2-\sqrt{2+\sqrt{2+\sqrt{3}}}},$$

the positive value of the root being taken in every case.

Show that, if $ax^3 + 3bx^2 + 3cx + d$ is divisible by $ax^2 + 2bx + c$ the former is a perfect cube, the latter a perfect square.

4. Solve the equations:

$$\text{(i.)} \quad \frac{1}{x} + \frac{1}{a} = \sqrt{\frac{1}{a^2} + \sqrt{\frac{1}{a^2x^2} + \frac{1}{x^4}}},$$

$$\text{(ii.)} \quad \begin{cases} 3x^2 + 6yz + 3(y+z) + 1 = 0, \\ 3y^2 + 6zx + 3(z+x) + 1 = 0, \\ 3z^2 + 6xy + 3(x+y) + 1 = 0. \end{cases}$$

In (ii.) account for the theoretical eight sets of solutions.

5. Two vessels, holding respectively 80 pints and 20 pints, are filled, the first with wine, the second with water. Two other vessels, of equal capacity p pints, are filled from the first and second of these respectively and then emptied into the vessel from which they were not filled. Find the composition of the mixture in each after two such operations; and show that, if the two liquids have then the same composition, their composition was the same after a single operation, and will remain so after any number of operations of the same kind. Find the value of p necessary for this result.

6. Find two numbers, each a perfect square of four digits, such that their difference is equal to the cube of the difference of their square roots.

Show how every perfect cube may be expressed as the difference of two perfect squares, and express thus the cube of 13.

7. When is a series of quantities said to be in Harmonical Progression? Find the nth term of such a progression, having given the first and third terms.

If $a-b$, $a-c$, $a-d$ are in Harmonical Progression, prove that the following three series of quantities are equally so:

(i.) $b-c,\quad b-d,\quad b-a,$

(ii.) $c-d,\quad c-a,\quad c-b,$

(iii.) $d-a,\quad d-b,\quad d-c.$

8. Find the total number of combinations which can be formed out of n different things.

My cousin belongs to a club of 30 members. In how many ways can I meet at least five of the members of the club, of whom my cousin may be one; and in how many ways can I meet five members or more, not including my cousin?

9. If $f(p)$ denote the series $1+p+\frac{p^2}{\underline{|2}}+\frac{p^3}{\underline{|3}}+\ldots$ to infinity, prove that $f(m)\times f(n)$ is identically equal to $f(m+n)$.

Hence deduce the series for the expansion of a^x in a series according to ascending powers of x.

10. Solve the equation

$$\left(\frac{295}{867}\right)^{3-x}=632\left(\frac{56}{99}\right)^{\frac{5x}{9}}$$

determining x correctly to 3 places of decimals.

IV. PLANE TRIGONOMETRY AND MENSURATION.

N.B.—*Great importance is attached to accuracy.*

1. Define *circular measure*, and the *sine of an obtuse angle*; and explain the meaning of the symbol π.

The circular measure of each of the angles of a regular figure of 77 sides is calculated to be 3·06; determine the assumed value of π.

2. Explain the *convention of signs* with reference to directions in a plane.

If $225° < A < 270°$, and $0° < B < 45°$, establish the identity

$$\cos(A-B) = \cos A \cos B + \sin A \sin B;$$

and show that the sign of $\cos(A-B)$ as found from the identity is the proper sign according to the convention.

3. Express $\cos 3A$ in terms of $\cos A$; and hence find the value of $\sin 18°$.

4. Find a general formula for all the values of θ which satisfy the equation $$3 \operatorname{cosec} \theta - 4 \sin \theta = 4.$$

5. If $A+B+C=2S$, prove that

$$\sin(B-C)\cos(S-A)+\sin(C-A)\cos(S-B)+\sin(A-B)\cos(S-C)=0;$$

and if $\alpha = \beta + \gamma$, show that

$$\sin(\alpha+\beta+\gamma)+\sin(\alpha+\beta-\gamma)+\sin(\alpha-\beta+\gamma)=4\sin\alpha\cos\beta\cos\gamma.$$

6. Prove that, in a plane triangle

$$a^2 = b^2 + c^2 - 2bc \cos A;$$

and adapt this formula to the logarithmic calculation of a by means of a subsidiary angle. If two sides and the included angle of a triangle be respectively 521, 479, and 59° 49′ 28″, find the third side.

7. A and B are two points on one bank of a straight river, distant from one another 649 yards; C is on the other bank, and the angles CAB, CBA are respectively 48° 31′ 10″ and 75° 24′ 50″; find the width of the river.

8. Prove that the diameter of a circle circumscribed to a triangle is equal to the product of one of the sides and the cosecant of the opposite angle.

The angle A of a triangle ABC, inscribed in a circle, is a radian; and the smaller arc BC is 20 ins.; find (by logarithms) the area of the smaller segment of the circle cut off by the chord BC. $(\pi = 3\frac{1}{7})$.

9. Prove that $\tan^{-1}\frac{5}{6} + \tan^{-1}\frac{7}{6} = \tan\ {}^{1}72$; and find x, when

$$\cos^{-1}\frac{1}{\sqrt{1+x^2}} - \cos^{-1}\frac{x}{\sqrt{1+x^2}} = \sin^{-1}\frac{1+x}{1+x^2}.$$

10. (i.) The slant side of a cone is 25 feet, and the area of its curved surface is 550 square feet; find its solid content. $(\pi = 3\frac{1}{7})$.

(ii.) If a sphere and a cube have equal surfaces, compare their volumes.

V. STATICS AND DYNAMICS.

[*The measure of the acceleration due to gravity may be taken to be* 32 *when the foot and second are the units of length and time.*]

1. Explain what is meant by the tension of a string and how it is measured.

A weight of 11 lbs. is suspended from a fixed point by a uniform string, which itself weighs 15 oz.; find the tension of the string at its extremities and at its points of trisection.

2. Find the resultant of two unequal forces acting on a rigid body in parallel but opposite directions.

AB is a heavy uniform rod, which is suspended from a fixed point by a string fastened to a point P of the rod, such that AP is one-third of AB. If the rod weighs 12 oz., find what weight, suspended by a string from the end A, will keep the rod horizontal. If the rod be then tilted round P into a different position in a vertical plane, will the equilibrium be disturbed?

3. Enunciate the proposition called the parallelogram of forces. Find the magnitude of the resultant of two forces equal to the weights of 51 lbs. and 68 lbs., acting at right angles to each other.

4. AB is a straight line and C is its middle point; find the centre of mass of three equal particles, placed at the points A, B, C.

If the particle at C be moved to a point D, such that ADB is an equilateral triangle, find the new position of the centre of mass.

5. Find the ratio of the Power to the Weight on a Wheel and Axle.

The radius of the wheel being three times that of the axle, find how far the weight will be lifted when the power is pulled down through the space of one foot.

6. Describe the common steelyard, and show that the distances between the graduations are proportional to the differences of the weights to which they belong.

7. Explain how velocity is measured. A railway train passes over 333 miles in 7 hours 24 minutes; find its average velocity in miles per hour and in feet per second.

8. Explain how uniform acceleration is measured.

If a stone is let fall, in what time will it reach a depth of 144 feet, and what will be its velocity at that time? If its mass is 8 oz., express its energy at that time in foot-pounds.

9. A heavy body is placed on a smooth inclined plane and allowed to slide down. The inclination of the plane to the vertical being two-thirds of a right angle, find the time of descent down a length of 128 feet of the plane, and the velocity acquired.

10. Two weights are connected by an inextensible string, which passes over a smooth fixed pulley; if the weights are held so that the portions of string below the pulley are vertical, and are then let go, determine the motion of each weight and the tension of the string.

If one of the weights is treble the other, find in what time the greater weight will descend through 32 feet, assuming that the string is long enough for this motion to take place.

11. Find expressions for the vertical and horizontal displacements at any time of a projectile *in vacuo*.

If the velocity of projection, in a direction inclined to the vertical at two-thirds of a right angle, is 64 feet per second, find the greatest height attained by the projectile. Also find the range of the projectile on the horizontal plane through the point of projection.

FURTHER EXAMINATION.

VI. PURE MATHEMATICS.

[*Full marks may be obtained for about* two-thirds *of this paper.*]

1. *A*, *B*, *C* are points in a straight line *ABC*; *P*, *Q*, *R* are the centres of squares described upon *AC*, *AB*, and *BC* respectively, *Q* and *R* being on the side of *AC* remote from *P*. Show that *BP* is equal to *QR*, and at right angles to it.

2. Given a point *O*, and three lengths *OA*, *OB*, *OC*, find the conditions under which the triangle *ABC* has a maximum area.

3. Write down the general term in the expansion of $(1-x)^n$.
Under what circumstances does the expansion (1) terminate, (2) converge, n being a real quantity?
Given that three consecutive coefficients are -20, 190, and -1140; find n.

4. A rectangle of perimeter b is inscribed in a circle of radius a. Find its sides and determine the limits of possibility of the problem.

5. Prove the formula

$$e^x = 1 + \frac{x}{1!} + \frac{x^2}{2!} + \frac{x^3}{3!} + \dots \textit{ ad inf.}$$

Show that
$$\sum_{n=1}^{n=\infty} \frac{n^3}{n!} = 5e.$$

6. Prove that

$$2\cos 2x \operatorname{cosec} 3x = \operatorname{cosec} x - \operatorname{cosec} 3x.$$

Sum to n terms the series

$$\cos 2x \operatorname{cosec} 3x + \cos(2 \,.\, 3 \,.\, x)\operatorname{cosec} 3^2x + \cos(2 \,.\, 3^2 \,.\, x)\operatorname{cosec} 3^2x + \dots.$$

7. From the top *A* of a cliff 597 feet high, the angle of elevation of a balloon *B* was observed to be 47° 20′ and the angle of depression of its shadow *S* upon the sea was 61°. *A*, *B*, and *S* being in the same vertical plane, and the altitude of the sun being 65° 31′, find the height of the balloon above the sea level.

8. In question 1, taking A as origin and AC as axis of x, find equations in rectangular co-ordinates to the lines BP, AR, CQ, QR, PQ, PR.

9. In a circle if any chord be drawn through a fixed point and tangents be drawn at its extremities, find the locus of their points of intersection.

Show that if one point lie on the polar of a second point, the second point will lie on the polar of the first point.

10. Find the polar equation of a parabola. Having given the lengths of two tangents to a parabola at right angles to one another, find the length of the latus rectum.

11. Express the length of the normal (terminated by the major axis) to an ellipse

$$\frac{x^2}{a^2}+\frac{y^2}{b^2}=1.$$

in terms of the inclination of the normal to the major axis.

Show that the sum of the squares of the normals drawn at the extremities of conjugate semi-diameters and terminated by the major axis is

$$a^2(e^2-1)(e^2-2).$$

12. Show that the locus of the vertex of a triangle constructed on a given base, one of whose base angles is double the other, is an hyperbola whose transverse axis is two-thirds of the base.

13. If a circle touch a given circle and a given straight line the locus of its centre is a parabola.

14. If PM be the perpendicular upon the directrix from any point P on an ellipse, then MS, drawn through the adjacent focus S, meets the normal at P on the minor axis.

VII. MECHANICS.

[*Full marks will be given for about* two-thirds *of this paper. Great importance is attached to accuracy. Gravitational acceleration may be taken* = 32 *feet per second per second, and* $\pi = \frac{22}{7}$.]

1. Investigate the conditions that three parallel forces, whose lines of action are intersected by a straight line ABC in A, B, C respectively, may be in equilibrium.

A solid cone of weight W, height h, and radius of base r, rests with its slant side in contact with a smooth horizontal plane. Find (i.) the least vertical force which, applied at the vertex, will move the cone, and (ii.) the force at the vertex which will keep it at rest with its axis horizontal.

2. Find the magnitude and direction of the resultant of a system of forces acting in one plane on a particle, the magnitudes and the directions of the forces being given.

Five forces P, Q, X, R, Y act at O, the centre of a regular pentagon $ABCDE$, in the directions OA, OB, OC, OD, OE respectively. If the system be in equilibrium and P, Q, R be given, find the magnitudes of X and Y. [N.B.—$\sin 18° = \frac{1}{4}(\sqrt{5} - 1.)$]

3. The sides AB and CD of a quadrilateral lamina $ABCD$ are parallel and at a distance h apart. If $AB = a$, $CD = b$, find the distance of the centre of mass of the lamina from the straight line bisecting the opposite sides AD and BC.

A rectangle $ABCD$ is bisected by a straight line cutting AB in E and CD in F. Show that the locus of the centre of mass G of either part is a parabola, and that the tangent at G to the locus is parallel to the corresponding line EF.

4. A rigid body is in equilibrium under the action of four forces in one plane, whose directions are not all parallel, and whose lines of action are not concurrent. If the magnitude of one of the forces be given, and the lines of action of all, show how the magnitudes of the other three may be found by the *graphical method.*

C and D are two small rough pegs, 1 foot apart in a straight line inclined to the horizon at an angle of 60°. A uniform rod $ACDB$ of 10 pounds weight and 3 feet in length is placed in this line, under C and above D, so that $AC = BD = 1$ foot. Find *graphically* the least weight which, suspended from the *upper* end B, will keep the rod in the above position, the angle of friction between the rod and the pegs being 45°.

5. Describe the common balance, and explain how its sensibility depends on the form of the beam.

6. Enunciate and explain the proposition known as the "Parallelogram of Velocities."

An insect crawls along the minute hand of a clock at the rate of 32 feet an hour. Find its velocity at the distance of one yard from the centre of the clock face.

7. A particle is projected up a rough inclined plane with velocity u. Find its position and velocity at the end of time t, the inclination of the plane to the horizon being i, and λ the angle of friction.

If the particle be projected with velocity due to falling from a point A above the point of projection, and through A a straight line be drawn at an angle λ to the horizon, show that this line will meet the inclined plane at the point where the particle comes to rest.

8. If a particle be moving in a vertical plane under the action of gravity, prove that its path must be a parabola whose axis is vertical, and that the velocity at any point varies as the normal (terminated by the axis).

Find the two directions in which a particle may be projected, with a velocity of 60 feet per second, so as to hit an elevated object whose horizontal and vertical distances from the point of projection are 48 and 20 feet respectively; and prove that the times of describing the two paths are as 4 to 13.

9. Two spheres of masses m_1 and m_2 impinge obliquely with velocities u_1 and u_2 in directions making angles a_1 and a_2 respectively with the line of impact; find their velocities and directions after impact, e being the coefficient of restitution.

Three equal billiard balls A, B, C are lying on a billiard table; A and B are in contact, and equidistant from C. The ball C, projected *very nearly* in the direction of the point of contact, strikes A first and B *immediately* afterwards. Show that after impact the velocities of A and B are in the ratio of 4 to $3 - e$.

10. Define work, potential energy and kinetic energy; and explain what is meant by the principle of the conservation of energy.

Three strings are knotted together at C. One string passes round a smooth peg A and supports a weight P at its free extremity; a second passes round a smooth peg B in the same horizontal line as A, and supports a weight Q; the third hangs vertically and supports a weight R. Prove that the work done in carrying C from its position of equilibrium to the peg A is

$$2Qc\sin^2\tfrac{1}{2}\theta$$

where $AB=c$, and θ is the angle which BC makes with the horizon in the position of equilibrium.

MATHEMATICAL EXAMINATION PAPERS

FOR ADMISSION INTO

Royal Military Academy, Woolwich,

NOVEMBER, 1897.

OBLIGATORY EXAMINATION.

I. EUCLID.

[*Ordinary abbreviations may be employed, but the method of proof must be geometrical. Proofs other than Euclid's must not violate Euclid's sequence of propositions. In the absence of special directions to candidates, any of the propositions within the limits prescribed for examination may be used in the solution of problems and riders. Great importance will be attached to accuracy.*]

1. Give definitions of *parallel straight lines*, *rectangle*, *sector of a circle*, *reciprocal figures*. State the enunciation of any proposition in the Sixth Book of Euclid in which *duplicate ratio* occurs, and give a numerical illustration.

2. Prove that if a straight line falls on two parallel straight lines it makes the alternate angles equal to one another.

3. To a given straight line, apply a parallelogram equal to a given triangle and having one of its angles equal to a given rectilineal angle.

4. Enunciate the two propositions which prove that, if P, Q, and R be three points in a straight line *in any order*, the sum of the squares on PR and QR differs from the square on PQ by twice a certain rectangle.

5. Divide a given straight line into two parts, so that the rectangle contained by the whole and one of the parts may be equal to the square on the other part.

6. Prove that if two circles meet they cannot have the same centre.

7. Prove that if, from the point of contact of a tangent to a circle, a chord be drawn, the angles which the chord makes with the tangent are equal to the angles which are in the alternate segments of the circle.

8. Prove that if two straight lines cut one another within a circle, the rectangle contained by the segments of one of them is equal to the rectangle contained by the segments of the other.

9. Inscribe in a given circle an equilateral and equiangular hexagon.

10. Two obtuse-angled triangles have one acute angle of the one equal to an angle of the other, and the sides about the other acute angle in each proportionals; prove that the triangles are similar.

11. Prove that if four straight lines are proportionals, the rectangle contained by the extremes is equal to the rectangle contained by the means.

12. Prove that in equal circles, sectors have the same ratio which the arcs on which they stand have to one another.

13. If the vertical angle of an isosceles triangle is half the angle of an equilateral triangle, show that the base is greater than half of one of the equal sides.

14. $ABCD$ is a quadrilateral in which ABC is a right angle; and the square on AD with twice the rectangle $AB \cdot CD$ is equal to the sum of the squares on AB, BC, and CD; show that the angle BCD is also a right angle.

15. A and B are fixed points on a circle whose centre is C, and P is a moving point on the circle; if AP revolves round A at a given rate, find the relative rates at which BP and CP revolve respectively round B and C.

16. $ABCDEF$ is a regular hexagon inscribed in a circle whose centre is O, and P is any point on the smaller arc AB; show that the perpendicular from P on OC is equal to the sum of the perpendiculars from P on OA and OB.

17. If in a quadrilateral $ABCD$, CD touches the circle through A, B, and C; and the rectangle $AB \cdot AC$ is equal to the rectangle $BC \cdot CD$; show that AB touches the circle through A, C, and D.

II. ARITHMETIC.

[*In order to obtain full marks arithmetical methods of solution are required, and all the working as well as the answers must be shown.*]

1. Find, in its lowest terms, the difference between $\frac{1001}{1365}$ and $\frac{935}{2295}$.

2. Find the product of $\cdot052 \times 1\cdot87 \times \cdot0021$, and divide the result by the product of $3\cdot5 \times 6\cdot63 \times 1\cdot10$.

3. Find the cost of carpeting a room 30 ft. long by 21 ft. wide with carpet 27 in. wide at 3*s*. 9*d*. per yard. How much would be saved by leaving 5 ft. at each end and 3½ ft. at each side uncarpeted?

4. What vulgar fraction of half-a-crown is equal to £$\cdot0\dot{1}\dot{2}$?

5. A man bought a horse for £37. 10*s*., and sold him so as to gain 25 per cent. At what price was the horse sold?

6. The interest for one year on £3060 after income tax at 6*d*. per £ is deducted from it amounts to £99. 9*s*. Find the rate per cent. of interest.

7. Find the value of 100,000 rupees in francs—one rupee being 1*s*. 3½*d*. and 25 francs 10 centimes being the equivalent of £1.

8. Show that the product of any three successive numbers is divisible by 6.

9. Upon a debt which is paid a year and a half before it is due, true discount of £11. 4*s*. is allowed, simple interest being reckoned at 4 per cent. per annum. What is the amount of the debt?

10. A cistern which is 9 ft. 4 in. long and 7 ft. 6 in. wide contains 6 tons 5 cwt. of water; if a cubic foot of water weigh 1000 ozs., what is the depth of water in the cistern?

11. If the surface of a cube be $491\cdot306406$ square inches, what is the length of its edge?

12. *A* works 7 hours a day for 6 days, *B* 4 hours a day for 9 days, and *C* 5 hours a day for 7 days: if the wages paid to *A* and *B* together amount to £7. 16*s*., what do *A*, *B*, and *C* each receive?

13. A man purchases a farm which on being let at £2 per acre pays him $3\frac{1}{8}$ per cent. on the investment; he purchases another which he lets for 2 guineas an acre, and it pays him $3\frac{3}{4}$ per cent. Compare the price which he paid per acre for the two farms.

14. A train overtakes two persons who are walking 2 miles and 4 miles an hour respectively, and completely passes them respectively in 9 seconds and 10 seconds: what is the length of the train and its speed in miles per hour?

III. ALGEBRA.

(Up to and including the Binomial Theorem, and the theory and use of Logarithms.)

[N.B.—*Great importance is attached to accuracy.*]

1. Explain carefully the meaning of the terms *algebraic sum* and *algebraic difference*, giving illustrations.

If $a^3 - 3a^2 - 2a + 1 = x - y$ and $a^3 + 3a^2 - 2a - 1 = x + y$ find x and y; and write down the product of $a^3 - 3a^2 - 2a + 1$ and $a^3 + 3a^2 - 2a - 1$.

2. (i.) Prove that the difference of the expressions $ax^3 + bx^2 + cx + d$ and $am^3 + bm^2 + cm + d$ is divisible by $x - m$, and write down the quotient.

What are the factors of

(ii.) $x^3 + y^3 + z^3$, having given that $x + y + z = 0$?

(iii.) $a^3(b-c)^3 + b^3(c-a)^3 + c^3(a-b)^3$?

3. Resolve into two factors each of the expressions

(i.) $x^4 + 1$ and

(ii.) $2x^2 + 7x + 2$,

calculating numerical coefficients correct to three places of decimals.

4. Simplify

(i.) $[a^{\frac{1}{3}} b^{\frac{1}{2}} (a^{-\frac{1}{6}} b^{-\frac{1}{4}})^{-\frac{1}{2}}]^{-\frac{4}{5}}$,

(ii.) $\sqrt{2\left(\frac{\sqrt{5}+\sqrt{3}}{\sqrt{5}-\sqrt{3}} + \frac{\sqrt{5}-\sqrt{3}}{\sqrt{5}+\sqrt{3}}\right)}$.

5. Solve the equations

(i.) $8(x-1) - 2\sqrt{(x-1)} - 15 = 0$.

(ii.) $x/a = y/b = z/c$ and $lx^2 + my^2 + nz^2 = 1$, x, y, and z being the unknowns.

In equation (i.) discuss whether either of the solutions you obtain is irrelevant, and, if so, what equation they satisfy.

6. If $a : b = c : d$, prove that

$$(ma + nc) : (mb + nd) = (na + mc) : (nb + md).$$

If $\frac{a}{b} > \frac{c}{d}$, prove that $\frac{ma+nc}{mb+nd}$ will be greater or less than $\frac{na+mc}{nb+md}$, according as m is greater or less than n, assuming all the letters to denote positive quantities.

7. If the first term of a geometric progression is a and the common ratio is r, prove the following results:

(i.) If $r < 1$ the sum to infinity is

$$\frac{a}{1-r}.$$

(ii.) If $r > 1$, and if the progression be continued *backwards* by placing terms *before* the term a, the sum to infinity of the terms so added is

$$-\frac{a}{1-r}.$$

8. Prove that the characteristic of the common (or Briggian) logarithm of a number N, when $N > 1$, is equal to the number of places that the first significant digit of N is to the left of the units' place, and state and prove a corresponding result when $N < 1$.

In the series $2+4+8+16+\ldots$ determine the number of digits in the 250th term, given $\log 2 = \cdot 30103$.

9. How many different arrangements may be made with the letters of the word *Llanfairpwllgwyngyll* taken all together?

How many of these arrangements begin and end with a *ll* and contain a third *ll* but not *four* consecutive *l*'s?

10. Prove the Binomial Theorem for a positive integral index, and extend it so as to embrace all *positive* values of the index.

IV. PLANE TRIGONOMETRY AND MENSURATION.

[N.B.—*Great importance will be attached to accuracy.*]

1. Define the cosine of an angle, and trace its changes in sign as the angle increases from zero to four right angles.

The angle C of the triangle ABC is equal to a right angle, and the sides AC, BC are respectively 10 and 20 feet. A perpendicular CD is drawn from C on AB, find the lengths of CD, AD, and BD.

2. Investigate expressions for the sine and secant of an angle when the tangent is known. What signs should be given to the radicals when the angle is (1) 200 degrees, and (2) 300 degrees?

Prove also that

$$\sin^4\theta + \sin^2\theta = 2 - 3\cos^2\theta + \cos^4\theta.$$

3. Prove that $\dfrac{\sin 2\theta}{2\sin\theta} = \cos\theta.$

Deduce from this (or otherwise determine) expressions for $\sin\frac{\theta}{2}$ and $\cos\frac{\theta}{2}$ when $\sin\theta$ is given. Apply the results to find $\sin\frac{\theta}{2}$ and $\cos\frac{\theta}{2}$ when $\sin\theta = -\frac{1}{3}$, and the angle θ lies between 270 and 360 degrees.

4. Investigate a general expression for all the angles whose sine is the same as $\sin\alpha$. Write down an expression for all the angles whose cosine is -1.

Find all the angles such that the tangent is twice the sine.

5. If $B + C = 180°$, prove that

$$2(1 - \sin B \sin C) = \cos^2 B + \cos^2 C.$$

If $A + B + C = 180°$, prove that

$$1 - 2\sin B \sin C \cos A + \cos^2 A = \cos^2 B + \cos^2 C,$$

and if $A + B + C = 0$

$$1 + 2\sin B \sin C \cos A + \cos^2 A = \cos^2 B + \cos^2 C.$$

6. Prove that in any triangle the sines of the angles are as the opposite sides.

The angles A, B of a triangle are respectively 40° 30′ and 45° 45′, and the intervening side is 6 feet, find the smaller of the remaining sides.

7. A tower standing on the edge of a cliff is viewed by a man lying down on the shore. The tower and the cliff immediately under the tower are found to subtend each an angle of 30°, while the distance of the eye from a point at the foot of that part of the cliff is 20 feet. Find the height of the tower.

8. The sides of a triangle being given, investigate an expression for the radius of the circumscribing circle.

Prove that in an equilateral triangle the radii of the inscribed, circumscribed, and escribed circles are as 1 : 2 : 3.

9. If $A+B-C=0$, prove that

$$\tan A \tan B \tan C = \tan C - \tan A - \tan B.$$

Find also the value of $\tan^{-1}x + \tan^{-1}\frac{1-x}{1+x}$.

10. Investigate an expression for the area of the curved surface of a right cone when the angle of the cone and the altitude are given.

It is required to divide a right cone into two portions by a plane parallel to the base, so that the surfaces of each portion (including the flat faces) may be equal. Find the ratio into which the slant sides are divided by the plane, and determine if the problem is always capable of solution.

V. STATICS AND DYNAMICS.

[*The measure of the acceleration due to gravity may be taken to be* 32 *when the foot and second are the units of length and time.*]

1. Explain why it is that forces can be represented geometrically by straight lines, and enunciate the propositions called the triangle of forces and the polygon of forces. What are the converses of these propositions? Are they necessarily true?

2. Find the resultant of two parallel forces acting in the same direction.

Two workmen are carrying a heavy ladder on their shoulders; find the portion of the weight of the ladder supported by each workman, the centre of gravity of the ladder being at the distance of three feet from one workman and of six feet from the other.

3. If three forces, acting in one plane, balance each other, prove that their lines of action are either concurrent or parallel.

Three rods, lying on a horizontal plane, are jointed together so as to form a triangle ABC, and a tightened string connects a point P in AB with a point Q in AC; find the directions of the stresses at B and C, and determine, by a graphic construction (hand-sketch), the direction of the stress at A.

4. Having given the positions of the centres of gravity of a solid body and of a given part of it, find the position of the centre of gravity of the remainder.

ABC is a triangular area, and D, E, F are the middle points of BC, CA, AB: if G is the centre of gravity of the triangle ABC, and if H is the centre of gravity of the quadrilateral $BFEC$, prove that HG is one-ninth of AD.

5. Neglecting the weights of the pulleys, find the ratio of the power P to the weight W in the system of pulleys in which the string which passes round any pulley has one extremity fixed and the other attached to the pulley next above it, the portions not in contact with any pulley being all parallel.

If there are three pulleys, each of weight w, find the relation between P, W, and w.

6. A man walking at the rate of three miles an hour on a straight road is passed at the same instant by two cyclists, riding in the direction in which he is walking, one at the rate of seven miles an hour and the other at the rate of eleven miles an hour; all going at the rates indicated for three minutes, what will then be the distances of the cyclists from the man walking?

7. If a body, starting with the velocity u, moves in a straight line with the uniform acceleration f, prove that the space s, passed over in the time t, is given by the equation

$$2s = 2ut + ft^2.$$

A body, moving with uniform acceleration in a straight line, is observed to pass over 24 feet in three seconds, and 42 feet in the three seconds following; find its acceleration.

8. A stone is projected vertically upwards with the velocity of 96 feet per second; neglecting the resistance of the air, find the height to which the stone would rise, and the time in which it would attain that height.

Also find the times at which the stone would be at the height of 128 feet above the point of projection.

9. Prove that the time of descent of a particle down any chord of a vertical circle beginning at the highest point of the circle is the same.

Find the line of quickest descent from a given point to a given straight line, the point and the line being in the same vertical plane.

10. A mass of 15 lbs. is held on a smooth horizontal table, at the distance of 4 feet from the edge of the table, and a fine string is attached to it, passing over the edge of the table and supporting a mass of one pound. If the 15 lb. mass be let go, find the velocity with which it will reach the edge of the table.

Describe the subsequent motion of the system.

FURTHER EXAMINATION.

VI. PURE MATHEMATICS.

[*Full marks may be obtained for about* two-thirds *of this paper.*]

1. The side AB of a triangle ABC is produced to D. If BE be drawn bisecting the angle CBD so as to meet AC produced in E, show that the square on BE is equal to the difference of the rectangles $AE \cdot EC$ and $AB \cdot BC$.

2. Draw a straight line which shall touch two given circles.

Two circles being given in position and magnitude; draw a straight line cutting them so that the chords in each circle may be equal to a given line, not greater than the diameter of the smaller circle.

3. Explain why, when quadratic equations are formed to express the conditions of a problem, the resulting roots may exceed in number what appear to be required as answers to the problem.

What does a merchant pay for brandy, when selling it for £39, he gains as much per cent. as the brandy cost him in sovereigns? Give an interpretation to the negative root which presents itself.

4. A steamer is running due west at 20 knots (miles per hour) in latitude such that its clock must be put back at the rate of one minute every 10 miles.

Find the run between consecutive noons, and the increase of speed necessary to make the run 520 miles.

5. Establish the formula

$$\frac{1}{2}\log_e y = m + \frac{1}{3}m^3 + \dots + \frac{1}{2n-1}m^{2n-1} + \dots, \text{ where } m = \frac{y-1}{y+1}.$$

Employ it to calculate $\log_e 10$ to four places of decimals, and thence calculate the multiplier which serves to convert the Napierian logarithms into ordinary or Briggian logarithms.

6. Show *a priori* that the radii of the circles circumscribed and inscribed to a triangle must be expressible as symmetric functions of the sides.

Determine such expressions.

7. A steamer running due west at 15 knots observes that a fixed point of an iceberg, which is drifting uniformly due south, bears $45°$ to the northward of the vessel's course. At intervals of five minutes the bearings are found to be $50° \; 5'$ and $56° \; 35'$ respectively.

Determine the rate of drifting of the iceberg.

8. Find the length of the perpendicular from the focus on a tangent to a parabola, $y^2 = 4ax$; the inclination of the tangent to the axis being $60°$.

Find the locus of the vertex of a parabola having a given point as focus and touching a given right line.

9. Find the equation of a straight line perpendicular to a given straight line.

A parallel being drawn to the base of a triangle so as to cut the sides; find the locus of the intersection of perpendiculars to the sides through the points of cutting.

10. Write down the general equation of a circle.

Find the locus of a point from which lines being drawn to several given points the sum of their squares shall be of given magnitude.

11. Find the equation of an equilateral hyperbola, taking the asymptotes as axes.

A straight line PCM is drawn from a fixed point P to intersect a fixed straight line AB in C and is produced to a point M such that a perpendicular MD to the fixed line intercepts a segment CD of given magnitude. Show that M is on an equilateral hyperbola, and find its centre and asymptotes.

12. Show that a curve of the second degree can be described in general so as to satisfy five conditions.

If you are given (1) the centre, (2) a system of conjugate diameters as regards direction and point of intersection; to how many conditions are these data respectively equivalent?

13. The exterior angle between any two intersecting tangents to a parabola is equal to the angle which either of them subtends at the focus.

Prove that if S be the focus and PQ the points in which two fixed tangents are cut by a third, the triangle SPQ will have its angles constant.

14. Prove that the straight line joining the foci of an ellipse subtends at the pole of any chord half the sum or difference of the angles which it subtends at the extremities of the chord. Distinguish between the cases.

VII. MECHANICS.

[*Full marks will be given for about* two-thirds *of this paper. Great importance is attached to accuracy. Gravitational acceleration may be taken* = 32 *feet per second per second, and* $\pi = \frac{22}{7}$.]

1. Show how to find the magnitude and line of action of the resultant of a system of coplanar forces.

Draw a straight line AB, 3 inches long, and upon it as base construct a triangle ABC whose sides CA, CB are $2\frac{1}{2}$ and $1\frac{1}{10}$ inches respectively. If forces of 5 lbs. act from C to A and from C to B, and a force of 6 lbs. from B to A, find *graphically* the magnitude and line of action of the resultant of the three forces.

2. If M_1, M_2 be the algebraical sums of the moments of a system of coplanar forces about two points A, B, prove that the algebraical sum about a point C lying between A and B will be

$$\frac{M_1 . BC + M_2 . AC}{AB};$$

and *hence* show that, if the algebraical sums of the moments of such a system with respect to three points not in a straight line, be separately equal to zero, the sum will be zero for any other point in the same plane.

3. The lower end B of a uniform rod AB, of length $2a$, rests on a smooth horizontal plane BD, and the upper end A is supported by a string AC of length b, the extremity C being fixed at a point whose height above the plane is $2a$. Find (i.) the horizontal force which, applied at the lower end of the rod, will cause it to rest at a distance b from the point in the plane immediately below C, (ii.) the reaction of the plane at B, and (iii.) the tension of the string AC;—the weight of the rod being W.

4. A uniform wire is bent into the form of a triangle; find the position of its centre of gravity.

If the lengths of the sides be 3, 5, and 7 inches, and the triangle be suspended from the obtuse angle, prove that in the position of equilibrium the shortest side will be horizontal.

5. State the laws of Limiting Friction, and explain what is meant by the "coefficient of friction" and the "angle of friction."

A lamina in the form of an equilateral triangle rests within a fixed rough hoop, the plane of each being vertical, and the radius of the hoop equal to the side of the triangle. Show that in the limiting position of equilibrium, the base of the triangle will be inclined to the horizon at an angle equal to twice the limiting angle of friction.

6. Prove that the work done in drawing a body up a rough inclined plane is equal to the work done in drawing the body along an equally rough horizontal plane through a distance equal to the length of the base of the inclined plane, together with the work done in lifting the body through the height of the plane.

A train whose mass is 96 tons commences the ascent of an incline of 1 in 80 at the rate of 45 miles an hour, the resistance being 7 lbs. per ton. If after travelling $\frac{1}{2}$ a mile the velocity be reduced to 30 miles per hour, find in foot tons the work done by the engine.

7. Find the tension of a light string which passes round a smooth fixed peg, and has unequal masses m_1, m_2 attached to its extremities.

Two bodies of equal mass hang from the ends of a light string which passes round a fixed peg, and when at rest are 9 feet and 16 feet above a fixed horizontal plane. A piece of the upper body breaks off and strikes the plane at the same moment that the lower body does. Find the ratio of the parts into which the upper body is divided.

8. Show that there are generally two directions in which a particle may be projected with given velocity from a given point O so as to pass through a point P.

If R be the difference between the two ranges on a horizontal plane through O, and H the difference between the two greatest heights attained by the particle in the two paths, prove that $R = 4H \tan\theta$, where θ is the angle OP makes with the horizon.

9. A smooth sphere impinges obliquely on a fixed plane; find the motion after impact, the elasticity being imperfect.

Two billiard balls P, Q of radius r stand upon a smooth table $ABDC$, their centres being at distances h_1, h_2 from the cushion AC and k_1, k_2 from the cushion AB. Find the direction in which P must be struck so that after impinging in succession upon AB, AC it may strike Q directly, taking e to be the coefficient of restitution of the indiarubber cushion.

10. The edges of a groove cut in a smooth horizontal table are two concentric circles whose radii are 11 and 13 inches. A sphere of 1 lb. mass whose diameter is 3 inches moves in this groove with uniform velocity. Find the greatest number of revolutions it may make per second without leaving the inner edge, and if it make half this number, find the reactions at the points where it rests upon the edges.

MATHEMATICAL EXAMINATION PAPERS

FOR ADMISSION INTO

Royal Military Academy, Woolwich,

JUNE, 1898.

OBLIGATORY EXAMINATION.

I. EUCLID.

[*Ordinary abbreviations may be employed, but the method of proof must be geometrical. Proofs other than Euclid's must not violate Euclid's sequence of propositions. In the absence of special directions to candidates, any of the propositions within the limits prescribed for examination may be used in the solution of problems and riders. Great importance will be attached to accuracy.*]

1. Enumerate (without proofs) the cases in which with certain data, two triangles are equal in every respect.

Show that the proof of Prop. 4, Book I., may require one of the triangles to be moved out of its plane.

2. Show that if a parallelogram and a triangle have the same base and altitude, the area of the first is double that of the second.

Hence prove that the area of a trapezium is half the sum of the parallel sides multiplied by the perpendicular distance between them.

3. Show that in any right-angled triangle, the square which is described on the side subtending the right angle is equal to the sum of the squares described on the sides which contain the right angle.

4. ABC is a triangle, and P the foot of the perpendicular from C on AB. Show that the square of AC differs from the sum of the squares of AB and BC by twice the rectangle under AB and BP.

Hence, if M is the middle point of AB, find the length CM in terms of the sides of ABC.

5. Prove any proposition in Book II. which shows that the rectangle under the sum and difference of two right lines is equal to the difference between their squares.

Hence give a purely geometrical construction for a point which divides a given line AB into two parts, the difference of whose squares is a given area.

6. Prove that if a circle passes through the vertices of a quadrilateral, the sum of a pair of opposite angles in the quadrilateral is two right angles.

Prove that if the sum of a pair of opposite angles in a quadrilateral is two right angles, a circle can be described round it.

7. Show how to construct an angle which is one-fifth of two right angles. What use is made of this by Euclid in Book IV.?

8. Show how to inscribe in a given circle a regular polygon of 15 sides.

9. Find in the base of a given triangle two points which divide the base into segments having to each other the ratio of the two sides (giving the proof).

A and B are two given points, and CD a right line parallel to AB; find on CD a point whose distance from A is three times its distance from B.

What is the greatest distance between AB and CD that will allow of a real solution?

10. Define similar triangles.

In a right-angled triangle, if a perpendicular be drawn from the right angle to the base, the triangles on each side of it are similar to the whole triangle and to one another.

11. Define a *mean proportional* between two magnitudes.

AB is a given right line; *O* is a given point on *AB* produced through *B*; find by a geometrical construction a point *P* on *AB* such that *PO* is a mean proportional between *AO* and *BO*.

12. In a given triangle *ABC* inscribe a rectangle, one of whose sides lies along *AB* and is twice the other side.

II. ARITHMETIC.

[*In order to obtain full marks arithmetical methods of solution are required.*]

1. What is the least fraction which, added to the sum of $\frac{29}{12}$ and $\frac{15}{16}$, will make the result an integer?

2. It requires 1344 tiles, each 9 in. by $4\frac{1}{2}$ in., to cover a court-yard. What is its area?

3. *A* and *B* own a field in shares proportioned as 15 to 13. If *A*'s share is $\frac{3}{8}$ of an acre, what is the size of the field in square yards?

4. What decimal of 11 cwt. is 3 qrs. 21 lbs.? *Or,*

What decimal of 10 kilograms is 75 grams?

5. Find the product of 0·0119 and 2·967. Divide this product by 21·93.

6. What amount of interest does £1020 yield per annum, if invested at $2\frac{3}{4}$ per cent.?

7. Find the least common multiple of 78, 84, and 90.

Three cyclists who are riding together have machines with wheels, respectively, 78, 84, and 90 inches in circumference; what is the least distance in yards that they must travel in order that their wheels shall be simultaneously in the same position as at starting?

8. Find the decimal equivalent to the fractional expression

$$\frac{(1-\frac{5}{13})\times 6\frac{1}{2}\times(\frac{1}{3}\div 2)}{\frac{4}{9}\times 11\frac{1}{4}}.$$

9. A box 4 feet long, 2 feet wide, and $1\frac{1}{2}$ feet deep (internal measurements) weighs 10 lbs. If it be filled with water weighing 1000 oz. for each cubic foot, what is the total weight of box and water?

10. *A*, *B*, and *C* go into partnership, *A* investing £4000, *B* £3000, and *C* £2000. If the profits for the first year are £1530, how must they be divided?

11. A man who has invested £6480 in $2\frac{1}{2}$ per cent. Consols at 108, sells out at 112, and invests the proceeds in 6 per cent. preference shares (nominal value £100), thereby increasing his income by £90. At what price did he buy the shares? (Neglect brokerage.)

12. Write down all the prime numbers between 50 and 100. Can you give any reasons why not more than two of these prime numbers are consecutive odd numbers?

13. A Parliamentary grant is made at the rate of 5*s*. per head for all the children at elementary schools. If this grant is distributed at the rate of 5*s*. 9*d*. per child in town and 3*s*. 3*d*. per child in country schools, what percentage of the total number of children are in each class of school?

14. A batsman has a certain average of runs for 16 innings. In the 17th innings he makes a score of 85 runs, thereby increasing his average by 3. What is his average after the 17th innings? (There are no "not out" innings.)

III. ALGEBRA.

(Up to and including the Binomial Theorem, and the theory and use of Logarithms.)

[N.B.—*Great importance is attached to accuracy.*]

1. Prove the truth of the following statements:

(i.) The sum of the cubes of three consecutive numbers is divisible by the sum of the numbers.

(ii.) The sum of the cubes of any three numbers diminished by three times the product of those numbers is divisible by the sum of the numbers.

2. Find the greatest common measure of $x^4-7x^3+15x^2-16x+21$ and $x^4+x^3-12x^2+x-3$.

Show that if a and b denote two numbers which have no common factor, then a^2+b^2 and a^2-b^2 have no common factor unless it be 2.

3. Express in their simplest forms

$$\text{(i.)}\quad \frac{a}{(a-b)(a-c)}+\frac{b}{(b-a)(b-c)}+\frac{c}{(c-a)(c-b)};$$

$$\text{(ii.)}\quad \frac{a^3(b-c)+b^3(c-a)+c^3(a-b)}{a^2(b-c)+b^2(c-a)+c^2(a-b)};$$

$$\text{(iii.)}\quad \frac{a+b}{a^3-b^3}+\frac{a-b}{a^3+b^3}-\frac{2(a^2-b^2)}{a^4+a^2b^2+b^4}.$$

4. Assuming that $x^m \times x^n = x^{m+n}$ for all values of m and n, find what meaning must be given to x^0 and $x^{\frac{p}{q}}$ where p and q denote positive integers.

Find the continued product of

$$\sqrt{\frac{a^{\frac{1}{3}}x^{\frac{1}{2}}}{b}},\quad \left(\frac{b^{\frac{1}{2}}x^{-\frac{1}{3}}}{y}\right)^{\frac{1}{2}},\quad \left(\frac{ax^{\frac{1}{3}}y^{\frac{1}{2}}}{b^{-\frac{1}{3}}}\right)^{\frac{3}{2}} \text{ and } \frac{1}{a(by)}.$$

5. Show that a quadratic equation cannot have more than two roots.

The difference of the roots of a quadratic equation is d, and the quotient obtained by dividing the greater root by the smaller is q. Find the quadratic equation.

6. Solve the equations

(i.) $\sqrt{ax+b}+\sqrt{ax+c}=\sqrt{2(b+c)}$;

(ii.) $3x-11y+7=5x+2y-8=11x-12$;

(iii.) $\dfrac{3x-1}{2x+7}+\dfrac{2x+7}{3x-1}=4\tfrac{13}{18}$.

7. If $\dfrac{bx+cy}{b+c-a}=\dfrac{cx+ay}{c+a-b}=\dfrac{ax+by}{a+b-c}$, and no numerator or denominator vanishes, show that either $x=y$ or $a=b=c$.

8. In how many different ways can the letters of the word *commission* be arranged when taken all together? In how many of these will the fourth and fifth places be occupied by the same letter?

I have two five pound notes, two sovereigns, two half-sovereigns, two florins, a shilling, and a sixpence; in how many ways can I give a donation?

9. By the aid of the binomial theorem, or otherwise, find the sum of the series

$$1+n+\frac{n(n-1)}{2!}+\frac{n(n-1)(n-2)}{3!}+\ldots,$$

n being an integer. (*N.B.*—2! denotes *factorial* 2.)

Find the term in the expansion of $\left(x+\dfrac{1}{x}\right)^{2n}$ which is independent of x.

10. Write down the expansions in powers of x of a^x and $\log_e(1+x)$ as far as x^5, and state any limitation which may be necessary for the validity of either expansion.

Find the coefficient of x^n in the expansion of $\log_e(1+x+x^2+x^3)$ in powers of x, distinguishing the various cases which may arise.

IV. PLANE TRIGONOMETRY AND MENSURATION.

[N.B.—*Great importance will be attached to accuracy.*]

1. Show how to connect the measure of an angle in degrees with the measure of the same angle in radians.

State (1) the numbers of degrees, and (2) the numbers of radians, in the angles described by the hands of a clock between 12 noon and 1.25 p.m. on the same day.

2. Define the sine of an angle, and prove that its value cannot exceed unity.

Write down the measures in degrees of the angles not exceeding two right angles whose sines are equal to 0, $\frac{1}{2}$, 1.

3. Obtain a formula for all the angles which have the same tangent as a given angle α.

If
$$\left.\begin{aligned} r\cos\theta &= a, \\ r\sin\theta &= b, \end{aligned}\right\}$$
and r, a, b are all positive, find a formula giving all the possible values of θ.

4. Prove geometrically the formula
$$\cos(A+B) = \cos A\cos B - \sin A\sin B$$
for the case where the angles A, B and $A+B$ are acute.

Prove that, if $\sin x = n\sin(x+2\alpha)$,

then
$$\tan(x+\alpha) = \frac{1+n}{1-n}\tan\alpha.$$

5. Prove that, if A, B, C are the angles of a triangle,

(i.) $\tan A + \tan B + \tan C = \tan A\tan B\tan C$;

(ii.) $\cot A + \cot B + \cot C = \operatorname{cosec} A\operatorname{cosec} B\operatorname{cosec} C + \cot A\cot B\cot C$.

6. Prove that the limit of $\frac{\sin\theta}{\theta}$ when θ is indefinitely diminished is unity.

Find approximately in seconds, the greatest angle than can be subtended by a yard measure at a distance of one mile.

7. Define the logarithm of a number to a given base, and show how to change from one base to another. Why may not 1 be taken as base?

Given $\log_e x = 2\pi$, find x from the tables.

[Assume $\log_{10}\pi = \cdot 4971499$ and $\log_{10}e = \cdot 4342942$.]

8. Prove that in any triangle ABC

$$a^2 = b^2 + c^2 - 2bc \cos A.$$

Prove that, if Δ is the area of a quadrilateral whose sides are of lengths a, b, c, d, and θ the sum of a pair of its opposite angles, then

$$16\Delta^2 = 2\{b^2c^2 + c^2a^2 + a^2b^2 + a^2d^2 + b^2d^2 + c^2d^2\} - (a^4 + b^4 + c^4 + d^4) - 8\,abcd \cos \theta.$$

9. Find formulæ for the solution of a triangle, given two angles and one side.

A man standing on one bank of a straight river sees two objects on the further side, and the lines joining his position to them make with the direction of flow of the river angles of 51° 36′ and 71° 48′. He walks down stream until the objects are seen in line, and finds that the line joining his position to them now makes an angle of 104° 57′ with the direction of flow of the river. He measures the distance he has walked, and finds it is 150 yards. What is the distance between the objects?

10. A cube standing upon a horizontal plane is cut by a plane which meets no vertical edge. Find expressions for the volumes cut off in the cases where (1) one of these edges is on one side of the plane and three on the other, (2) two of these edges are on one side of the plane and two on the other.

V. STATICS AND DYNAMICS.

[*The measure of the acceleration due to gravity may be taken to be* 32 *when the foot and second are the units of length and time.*]

1. Explain what is meant by the resolved part of a force in a given direction.

If the resolved part is half the force, what is the angle between the given direction and that of the force?

2. Find the resultant of two parallel forces acting in opposite directions on a rigid body.

A heavy uniform rod ACB, of weight w, is supported, in a horizontal position, by a vertical string fastened to its middle point C; from the end B a weight w is suspended; and from a point D in CA a weight Q is suspended. Find Q, and the position of the point D, when the tension of the string at the middle point is equal to $4w$.

3. Find the position of the centre of gravity of a uniform triangular lamina.

In the sides AB, AC of a triangle ABC, points D, E are taken, such that AD is one-third of AB, and AE one-third of AC, and the points D, E are joined by a straight line; if F is the middle point of BC, and if G and H are the centres of gravity of ABC and of the quadrilateral $BDEC$, find the ratio of HG to AF.

4. Describe the common steelyard, and show how to graduate it.

If the length of the steelyard is four feet, if the resultant of the weights of the steelyard and scale-pan, which is equal to three pounds, acts in a vertical line distant sixteen inches from the end to which the scale-pan is attached, and if nineteen pounds is the greatest weight which can be measured when the movable weight is three pounds, find the position of the fulcrum.

5. Find the ratio of the power to the weight in the wheel and axle.

If two masses, each of weight w, are fastened to points on the wheel at an angular distance from each other of 120°, find the position of the wheel when the greatest possible weight is supported on the axle, and find what is this greatest weight.

6. Explain in what sense a moving point may be said to have two velocities at the same time, and prove the rule for their composition.

A man in a railway train, travelling at a given rate, carries a loaded rifle, and fires it off so as to strike a telegraph post as he passes it; having given the velocity of the bullet, determine, by graphic construction, the direction in which he must aim.

7. Explain how uniform acceleration is measured.

How would the measure of the acceleration due to gravity be altered if the unit of time were half a second?

8. Two scale-pans, each weighing two ounces, are connected by a fine string passing over a smooth fixed pulley, and are hanging at rest. A weight of 15 oz. is placed in one scale-pan, and, simultaneously, a weight of 13 oz. is placed in the other scale-pan; find the acceleration of each scale-pan, the tension of the string, and the pressures of the weights on the scale-pans.

9. A stone is projected horizontally from the top of a tower, 256 feet in height, with the velocity of 128 feet per second; find when, and where, it will strike the horizontal plane on which the tower is situated, and the direction of its motion at that time.

10. If a point moves uniformly with velocity v in a circle of radius r, prove that its acceleration is in the direction of the centre of the circle, and is equal to v^2/r.

A heavy particle is attached to one end of a string three feet in length, the other is fastened to a fixed point; and the particle is held so that the string is inclined at the angle of 60° to the vertical. Find the velocity with which it must be projected horizontally so as to describe, uniformly, a horizontal circle.

FURTHER EXAMINATION.

VI. PURE MATHEMATICS.

[*Full marks may be obtained for about* two-thirds *of this paper.*]

1. The perpendiculars drawn from the angular points A, C of the triangle ABC, to the opposite sides, cut these sides in D and F, and intersect one another in P. Prove that the tangents, at D and F, of the circle which passes through $BDPF$, pass through the middle point of the side CA; and that the tangents of this circle at P and B are parallel to CA.

2. If one side of a triangle be twice another side of this triangle, prove that the angle opposite to the former side is greater than twice the angle opposite to the latter side.

3. Prove that

$$(a^2+b^2+c^2)(a'^2+b'^2+c'^2)=(aa'+bb'+cc')^2+(bc'-b'c)^2+(ca'-c'a)^2+(ab'-a'b)^2.$$

4. What is the condition for the validity of the binomial expansion of $(1+x)^n$ in ascending powers of x?

Find the sum of the series

$$1+\frac{3!}{(1!)^2}\cdot\frac{4}{25}+\frac{5!}{(2!)^2}\left(\frac{4}{25}\right)^2+\frac{7!}{(3!)^2}\left(\frac{4}{25}\right)^3+\ldots \text{ to } \infty.$$

5. Prove that the amount of a sum of £A, in n years, at r per cent. per annum compound interest, payable yearly is

$$A\left(1+\frac{r}{100}\right)^n.$$

Hence find, with the help of the table of logarithms, the time in which a sum of money will double itself, at $2\frac{1}{2}$ per cent. per annum compound interest, payable quarterly.

6. Two persons, P, Q, stationed on a coast which runs East and West, observe a ship when it is due North of P, and again, when it is due North of Q. In the former case it is 45° West of North as seen from Q, in the latter case it is 30° East of North as seen from P. Determine the direction which the ship is travelling.

7. Mark the points (2, 2)(− 2, 2)(− 2, 0) with reference to the rectangular axes, OX, OY, and find the area of the triangle joining them.

8. Prove that the equation $ax+by=c$ represents a straight line.

Find the point of intersection of $3x+2y=10$ and $2x+3y=10$, and write down the equations of the straight lines which pass through this point and are each equally inclined to the two rectangular axes.

9. Find the equation of the chord of the circle $x^2+y^2=a^2$ which joins the points (x', y'), (x'', y''); and deduce the equation of the tangent of the circle at any point.

Show that the common tangents of the circles

$$x^2+y^2+2x=0, \qquad x^2+y^2-6x=0$$

form an equilateral triangle.

10. The ends A, B of a straight line AB, of constant length a, slide upon the fixed rectangular axes OX, OY respectively. If the rectangle $OAPB$ be completed, show that the coordinates of the foot of the perpendicular drawn from P to AB are $a\cos^3\theta$, $a\sin^3\theta$, where θ is the acute angle AB makes with OX.

11. Prove that if two parabolas have the same focus and axis, the locus of the point of intersection of two tangents at right angles, one to each of the parabolas, is a straight line.

12. Prove that if the pole of one straight line in regard to a conic lie upon another straight line, then the pole of the latter lies on the former.

If the straight lines intersect outside the conic, prove that they are harmonic in regard to the tangents of the conic drawn from their point of intersection.

13. Show how to draw tangents to a parabola from any exterior point.

Given the focus and a point of a parabola, and the tangent at the point, show how to describe the parabola.

14. Determine the radius of curvature at any point of an ellipse.

Tangents PQ, PR are drawn to an ellipse from a point P. Find the limit of the radius of the circle PQR as the point P approaches indefinitely near to, and ultimately lies upon, the ellipse.

VII. MECHANICS.

[*Full marks will be given for about* two-thirds *of this paper. Great importance is attached to accuracy. Gravitational acceleration may be taken* = 32 *feet per second per second, and* $\pi = \frac{22}{7}$.]

1. If the directions and magnitudes of any forces in one plane be graphically represented by the sides AB, BC, CD, etc. of an unclosed polygon taken in order, prove that the direction and magnitude of the resultant is represented by the straight line closing the polygon.

If the polygon is closed, and the sides are the lines of action of the forces, determine whether the forces are in equilibrium.

A uniform heavy straight rod AB of one pound weight is placed with one end A resting on a rough horizontal plane and the other B against a smooth vertical wall. Supposing the rod to be on the point of motion, and the coefficient of friction to be $\frac{1}{2}$, determine by a graphical construction the magnitudes of the reactions at A and B, and the inclination of the rod to the horizon, giving the numerical results.

2. Prove that the sum of the moments of any two forces about any point in the plane of the forces is equal to the moment of their resultant about the same point.

Three forces P, Q, R act along the sides BC, CA, AB of a triangle ABC. Prove that their resultant passes through the orthocentre if

$$\frac{P}{\cos A}+\frac{Q}{\cos B}+\frac{R}{\cos C}=0.$$

3. Show how to find the centre of mass (centre of gravity) of a quadrilateral area, and prove that three times the distance of that point from any straight line in the plane of the area is equal to the sum of the distances of the four corners minus that of the intersection of the diagonals from the same straight line.

A heavy quadrilateral area $ABCD$ is suspended from one corner A, M is the middle point of the diagonal BD, and O is the intersection of the diagonals. Prove that in equilibrium the vertical through A divides MO in V, so that

$$\frac{MV}{VO}=\frac{1}{2}\frac{CO}{OA}.$$

4. A straight rod AB, whose weight can be neglected, is placed with its middle point C at the highest point of a fixed perfectly rough horizontal circular cylinder, the rod being at right angles to the axis. If two weights W, W' are suspended from the ends A, B, prove that the rod in its new position of equilibrium will be inclined to the horizontal at an angle

$$\frac{l}{a}\,\frac{W \sim W'}{W+W'},$$

where $2l$ is the length of the rod and a the radius of the cylinder.

How is the problem altered if the cylinder be not perfectly rough?

5. A system of n equal heavy pulleys, in which a separate string passes round each pulley with one end attached to a fixed horizontal beam and the other to the pulley next above (all the strings being vertical), has a power P acting upwards and a weight W downwards. Find the mechanical advantage.

Prove that the magnitude of the pull on the horizontal beam is $(P-w)(2^n-1)+nw$, where w is the weight of any pulley.

When there are five pulleys, prove that the distance of the point of application of the pull from the string of greatest tension is

$$\frac{26P-16w}{31P-26w}a,$$

where a is the radius of any pulley.

6. Particles whose masses are m_1, m_2, etc. are simultaneously projected on a smooth horizontal table with velocities V_1, V_2, etc. from a point O in directions making angles θ_1, θ_2, etc. with a fixed straight line Ox. Prove that their centre of gravity describes a straight line making an angle θ' with Ox where

$$\tan\theta' = \frac{\Sigma m V \sin\theta}{\Sigma m V \cos\theta}.$$

If two of the particles are connected together by a weightless slack string, determine how the motion of the centre of gravity is affected when the string becomes tight.

7. Find the time of descent from rest of a heavy particle on a smooth inclined plane.

Two heavy particles whose weights are W and W', slide down two smooth rods CA, CB in the same vertical plane, starting from rest at C. The rods are rigidly connected at C, which is fixed in space, and make angles θ and θ' with the vertical, on opposite sides of it. Prove that if the rods are held at rest, the centre of the circle circumscribing the particles and the point C moves vertically.

Assuming the rods to have no weight, prove that no force, except that fixing C, is necessary to hold them at rest as the particles descend if

$$W \sin 2\theta = W' \sin 2\theta'.$$

8. Prove that the path of a heavy projectile *in vacuo* is a parabola, and that the square of the velocity varies as the distance of the particle from the focus.

Two particles are simultaneously projected from the same point A with equal velocities in directions AT, AT' where AT makes the lesser angle with the vertical, and the parabolas intersect in B. Determine which particle arrives first at B.

Prove also that the product of their times of transit from A to B is equal to the square of the time occupied by a particle falling from rest vertically through a distance equal to AB.

9. Two smooth imperfectly elastic spheres whose masses are m, m', impinge directly on each other with velocities V, V'. Find the velocities after impact.

Two smooth spheres moving in directions at right angles to each other impinge obliquely. If their directions of motion after impact are also at right angles, prove that the coefficient of elasticity is

$$\frac{m^2v - m'^2v'}{mm'(v - v')},$$

where m, m' are the masses and where v, v' are the components of the velocities of the spheres in the direction of the straight line joining the centres just before impact.

10. Explain what is meant by the work of a force. What is horse-power?

A given weight W is hoisted up to a given height h by a workman. During the first third of the height the weight is made to move upwards with a constant acceleration, during the second third with a constant velocity, during the last third it is brought gradually to rest with a constant retardation. Supposing that the resistances (including gravity) are represented by a given constant retarding force R, and that the whole time of the ascent is equal to that of falling from rest through a height nh, prove that the difference of the works done by the man in the first and third portions of the ascent is

$$\frac{25}{18}\frac{Wh}{n}.$$

State also the whole work done.

MATHEMATICAL EXAMINATION PAPERS

FOR ADMISSION INTO

Royal Military Academy, Woolwich,

NOVEMBER, 1898.

I. PART I. FIRST PAPER.

[*Great importance will be attached to accuracy.*]

1. Reduce to its simplest form

$$\frac{(8\tfrac{4}{5}+\tfrac{1}{9}-\tfrac{1}{45})(4\tfrac{1}{6}-3\tfrac{1}{4})}{(1\tfrac{5}{11}+\tfrac{1}{8})-(\tfrac{9}{16}-\tfrac{1}{8}-\tfrac{1}{22})}.$$

2. Find the value of £1·875 − 2·25 crowns + 3·75 guineas − 46·125 shillings.

3. At what rate of simple interest must £230,398. 6*s.* 8*d.* be invested in order that it may amount to £298,941. 16*s.* 9*d.* at the end of seven years?

4. Of a certain store of potatoes, 11 men would in 3 days consume all but 201 pounds, and 21 men would in 4 days consume all but 48 pounds; how many pounds of potatoes does the store contain?

5. If an integer which is less than 50 become a square number when 25 is added to it, show that it can be resolved into two factors whose difference is 10, and hence show that in such an integer the units digit is equal to the square of the tens digit; write down all such integers.

6. Find the value of

$$\frac{3(x+y+z)(yz+zx+xy)-x^3-y^3-z^3}{x^2+y^2+z^2-yz-zx-xy}$$

when $x=\frac{1}{2}$, $y=\frac{2}{3}$, $z=\frac{3}{4}$.

7. Prove that the remainder which is left when x^4+7x+3 is divided by $x-1$ is the same as the remainder which is left when x^4-6x+7 is divided by $x-2$.

8. In any division sum show that a common factor of the divisor and the dividend is a factor of any of the successive complete remainders.

In finding the highest common factor of two expressions, $2x^4 - x^3 - 3$ was one of the divisors, and $5(x^3 + 3x + 4)$ was the corresponding remainder. What was the highest common factor of the two expressions?

9. Solve the equations

(i.) $\dfrac{(a^2-1)(ax+1)}{a^2(x+a)} + \dfrac{(a^2+1)(x-a)}{ax+1} = \dfrac{ax+1}{x+a} + \dfrac{a(ax-1)}{ax+1}$;

(ii.) $\begin{cases} \frac{3}{2}(x-y) + \frac{1}{8}(x+2y) = 4 \\ 2(x-y) + 3(x+2y) = 19; \end{cases}$

(iii.) $\dfrac{7}{x-1} + \dfrac{5}{x-2} + \dfrac{22}{3}\dfrac{1}{x-3} = 0.$

10. The mth and n^{th} terms of an arithmetical progression are p, q. Find the first term and the common difference.

The thirty-first term of a certain arithmetical progression exceeds the square of one-half of the fourth term by unity, and also exceeds twice the fourteenth term by unity; find its values.

11. Prove that the number of combinations of n things taken r at a time is

$$\frac{n!}{(n-r)!r!}.$$

You have seven envelopes directed to seven people, to four of whom you intend to send copies of a circular, the other three envelopes to be used for a different purpose. In how many incorrect ways can you put the four circulars in four of the seven envelopes?

12. Assuming that the binomial theorem is true for a positive integral exponent, prove that it is true for a positive fractional exponent.

Apply the binomial theorem to find the fifth root of 3126 to seven places of decimals.

II. PART I. Second Paper.

N.B.—*In questions on Geometry ordinary abbreviations may be employed, but the method of proof must be geometrical. Proofs other than Euclid's must not violate Euclid's sequence of propositions. In the absence of special directions to Candidates, any of the propositions within the limits prescribed for Examination may be used in the solution of problems and riders.*

1. Show that, if from the ends of the base of a triangle two right lines are drawn to any point within the triangle, the sum of these is less than the sum of the other two sides of the triangle, but they contain a greater angle.

Show that, if any point inside a triangle is joined to the three vertices, the sum of the joining lines is less than the sum, and greater than half the sum, of the three sides of the triangle.

2. Show that, if the square on one side of a triangle is equal to the sum of the squares on the other two sides, the triangle is right-angled.

Explain the meaning of the *converse* of a theorem. What is the converse of the theorem in this question?

3. Assuming Propositions 12 and 13 of Euclid, Bk. II., prove that the sum of the squares on the diagonals of any parallelogram is equal to the sum of the squares on the sides.

4. Prove that all angles contained in the same segment of a circle are equal, those in a semicircle being right angles.

If any two circles be drawn each touching the three sides of a triangle, prove that the circle described on the line joining their centres as diameter passes through two vertices of the triangle.

5. Show how to draw a tangent to a circle from any given external point.

Two points are given; one is to be the centre of a circle, and the tangent drawn to this circle from the other is to be of given length less than the distance between the given points; show how to draw the circle and the tangent.

6. Show how to describe a square about a given circle.

Every parallelogram described about a circle must have all its sides equal.

7. Define similar triangles, and state (without proof) the relation between their areas.

Give any method by which an equilateral triangle may be inscribed in a given triangle *ABC*, having one of its sides perpendicular to *AB*.

8. There is a piece of ground in the form of a trapezium, the lengths of the parallel sides of which are 20 and 34 yards, and the lengths of the other two sides 15 and 13 yards; find its area.

9. Prove that the area of the curved surface of a frustum of a cone is equal to the slant height multiplied by the perimeter of the mid section.

10. Find the number of cubic feet of earth removed per yard length from a cutting in level ground 12 feet in depth, the breadth of the base of the cutting being 15 feet, and the slopes of both sides 45°.

III. PART I. Third Paper.

1. Prove that the angle subtended at the centre of a circle by an arc which is equal in length to the radius is an invariable angle, and explain what is meant by the circular measure of an angle.

A circular wire of 3 inches radius is cut, and then bent so as to lie along the circumference of a hoop whose radius is 4 feet. Find the angle which it subtends at the centre of the hoop.

2. Define the *cotangent* and *cosecant* of an angle, and show that

$$\cot^2 A + 1 = \operatorname{cosec}^2 A.$$

Which is greater, the acute angle whose cotangent is $\frac{4}{5}$, or the acute angle whose cosecant is $\frac{5}{4}$?

3. Obtain an expression for all the angles which have a given tangent.

Find all the angles, lying between $-360°$ and $+360°$, which satisfy the equation

$$\tan^2 x - \frac{2}{\sqrt{3}} \tan x - 1 = 0.$$

4. Prove geometrically, for the case in which A and B are two positive angles whose sum is less than a right angle, that

$$\sin(A+B) = \sin A \cos B + \cos A \sin B.$$

Express $\dfrac{\sin 3A}{\sin 2A - \sin A}$ in terms of $\cos A$.

5. Prove the formulæ

$$(1+\cos A)\tan^2 \tfrac{1}{2}A = 1 - \cos A$$

$$(\sec A + 2\sin A)(\operatorname{cosec} A - 2\cos A) = 2\cos 2A \cot 2A.$$

6. In any triangle ABC, show that

$$\sin \tfrac{1}{2}A = \sqrt{\frac{(s-b)(s-c)}{bc}}.$$

Find the greatest angle of the triangle whose sides are 184, 425, and 541.

7. Define the logarithm of a number, and point out the use of logarithms in arithmetical calculations.

From
$$\log 2 = 0{\cdot}3010300$$
$$\log 3 = 0{\cdot}4771213$$
$$\log 11 = 1{\cdot}0413927$$

find the values of $\log 7{\cdot}92$, $\log 0{\cdot}001089$, and $\log \sqrt{\tfrac{11}{12}}$.

8. If ABC be a triangle, and θ such an angle that $\sin\theta = \dfrac{2\sqrt{ab}}{a+b}\cos\tfrac{1}{2}C$, find c in terms of a, b, and θ.

If $a = 11$, $b = 25$, and $C = 106^\circ\ 15'\ 37''$, find c, having given

$L\cos 53^\circ\ 7'\ 48''{\cdot}5 = 9{\cdot}7781509$;
$L\sin 33^\circ\ 33' = 9{\cdot}7424616$. Tab. Diff. 1904.
$L\cos 33^\circ\ 33' = 9{\cdot}9208555$. Tab. Diff. 838.

For other logarithms see Question 7.

9. If tangents be drawn to the inscribed circle of a triangle parallel to the sides of the triangle, show that the areas of the triangles cut off by these tangents are inversely proportional to the areas of the corresponding escribed circles.

10. Prove that

$$\log_e(x+1) - \log_e x = 2\left\{\frac{1}{2x+1} + \frac{1}{3(2x+1)^3} + \frac{1}{5(2x+1)^5} + \dots\right\},$$

and deduce that

$$\log_e 13 = 2\log_e 2 + \log_e 3 + {\cdot}0800427\dots.$$

IV. PART II. FIRST PAPER.

[*Full marks may be obtained for about* three-quarters *of this paper.*]

1. Solve the simultaneous quadratics

$$\text{(i.)} \quad \begin{cases} x^2 + xy = 10, \\ 2xy - y^2 = 3; \end{cases}$$

$$\text{(ii.)} \quad \begin{cases} x^2 + y^2 = 13, \\ 2x - xy + 2y = 4. \end{cases}$$

Give the general rule for solving when the unknown quantities occur symmetrically in the equations as in (ii.).

2. Give the mth term in the expansion of

$$(1 - x)^{\frac{1}{n}}$$

by the binomial theorem.

Prove that

$$\frac{1}{2}\sqrt[5]{28} = \left(1 - \frac{1}{8}\right)^{\frac{1}{5}} = 1 - \frac{1}{40} - \sum_{m=2}^{m=\infty} \frac{1 \cdot 4 \cdot 9 \ldots (5m-6)}{\lfloor m} \left(\frac{1}{40}\right)^m.$$

3. Prove the identity

$$\sec 2A = 1 + \tan A \tan 2A.$$

Solve the equation

$$\sin(\alpha + x) + \sin(\beta + x) = 0.$$

4. Find the area of a regular quindecagon inscribed in a circle of one foot radius, making use of the table of natural sines.

5. Inscribe an equilateral and equiangular pentagon in a given circle.

In the isosceles triangle *ABC*, whose base is *BC*, each base angle is double of the third angle. In *AB* take a point *D* such that *CD* bisects the angle *ACB*. Prove that *BC* is equal to the side of a regular pentagon inscribed in the circle *ADC*.

6. Prove that if a straight line be drawn parallel to one of the sides of a triangle it cuts the other sides or those sides produced proportionally.

In the side *BC* of a triangle *ABC* take a point *D* and draw *DE*, *DF* parallel to *CA* and *BA* respectively to meet *AB* in *E* and *AC* in *F*. Find the locus of the middle point of *EF*.

7. Show that, if a solid angle be contained by three plane angles, any two of them are greater than the third.

8. Prove that similar polygons inscribed in circles are to one another as the squares on the diameters of the circles.

9. Define the latus rectum of a conic section, and show that the tangents to the curve at its extremities intersect in the directrix.

Prove that the focal distance of a point on a conic is equal to the length of the ordinate produced to meet the tangent at the end of the latus rectum.

10. In a conic, SP is the focal distance of any point P, and PG is the normal. Prove that the ratio of SG to SP is constant.

Prove that if normals be drawn at the ends of a focal chord, a line through their intersection parallel to the transverse axis will bisect the chord.

11. P is any point on a conic, and T any point in the tangent at P. If TM be drawn perpendicular to the focal distance SP and TN perpendicular to the directrix, show that SM is to TN in a constant ratio.

PQ is any chord of a conic, subject to the condition that it subtends a constant angle at the focus, and the tangents at P and Q intersect at T. Show that the locus of T is a conic with the same focus and directrix.

12. If the straight line joining two points P, P' of a conic meet a directrix in F, and F be joined to the corresponding focus S, prove that FS will bisect one of the angles between PS and $P'S$.

V. PART II. SECOND PAPER.

[*Full marks may be obtained for about* three-quarters *of this paper. Great importance is attached to accuracy. Gravitational acceleration may be taken* = 32 *feet per second per second.*]

1. Obtain the components of a given force parallel and perpendicular to a given line.

A body which weighs 12 lbs. is kept at rest by means of two cords, one being horizontal, and the other inclined to the horizontal at an angle whose tangent is $\frac{3}{4}$; find their tensions.

2. On the two smooth fixed rods AB and AC, equally inclined to the vertical at an angle of 30°, slide two small rings each of weight 2 ounces. Another weight of 10 ounces is knotted at E to the light thread DEF, length 4·6 inches, and the ends D and F are tied to the rings. Find graphically the directions of the two parts of the thread, and the position in which it hangs between the rods.

Given $DE : EF = 1 : 2$.

3. Define the moment of a force about a point, explaining the rule of signs.

Prove that the sum of the moments of two parallel forces acting on a rigid body about any point in their plane is equal to the moment of their resultant about the same point.

4. Define the centre of mass of a body.

A uniform lever of weight 6 lbs. and length 18 inches has weights of 11 lbs. and 7 lbs. attached at its ends; find the centre of mass of the system.

5. State the conditions of equilibrium of a system of forces acting in one plane on a rigid body.

A rigid uniform bar of weight 10 lbs. rests with its lower end in contact with a horizontal rough plane and makes an angle of 30° with the horizontal, being kept in position by a horizontal thread attached to its upper end. Find the tension of the thread and the magnitude and direction of the force exerted at the lower end.

6. A body of given weight W lbs. rests on a smooth plane of inclination a to the horizontal under a force acting up the plane. Find the magnitude of the force and the pressure exerted by the plane on the body.

If the body is a uniform cube with four of its edges horizontal and one face in contact with the plane, and if the force applied to support it acts in a line of greatest slope in that face, find the least inclination of the plane for which the cube will turn about its lowest edge.

7. A pistol shot is fired from a railway carriage travelling with velocity u so as to strike an object seen (at the instant of firing) in a line at right angles to the direction of motion of the carriage. Assuming that the charge of powder in the cartridge can impart a velocity v to the shot, find the direction in which the pistol must be aimed.

8. Define the acceleration of a point moving in a straight line.

Prove that if the point moves with uniform acceleration its average velocity in any interval of time is half the sum of its velocities at the beginning and end of the interval.

Two particles move from the same point A along the same line AB, one of them having a uniform velocity u, and the other a uniform acceleration f and no initial velocity. Find the time that elapses before the second overtakes the first.

9. A body weighing W lbs. is carried up in a lift which moves with an acceleration of 4 ft. per sec. per sec. State the direction of the force exerted upon it by the floor of the lift, and the weight of a body which this force could support at rest.

How would the result be modified if the lift ascended with uniform velocity?

10. A body weighing W lbs. is acted on by a force which could support a weight of W' lbs. at rest. In what time will it acquire a velocity v feet per sec.?

A body is suspended by a cord from the roof of a railway carriage forming part of a starting train, and it is observed that the cord is inclined to the vertical at an angle whose tangent is $\frac{1}{8}$. In what time will the train acquire a velocity of 30 miles an hour?

11. Determine the magnitude and direction of the acceleration of a point describing a circle of radius a with uniform speed v.

A wheel of radius r is rolling uniformly with velocity v along a level road. State the magnitude and direction of the velocity and acceleration (*a*) of the point instantaneously highest, (*b*) of the point instantaneously lowest.

12. A shot is fired from the edge of a vertical cliff of height b to strike an object at a distance a from the foot of the cliff. Prove that, if the velocity of projection is that due to falling through a height h, the angle of elevation α is given by the equation

$$a^2 \tan^2\alpha - 4\,h\,a \tan\alpha + a^2 - 4\,h\,b = 0,$$

and hence find the condition that the object may be within range. The resistance of the air is to be neglected.

VI. PART III. FIRST PAPER.

[*Full marks will be given for about* two-thirds *of this paper.*]

1. Prove geometrically that the subnormal in the parabola is equal to the semi-latus rectum.

PQ is a chord perpendicular to the axis, and *QR* is drawn perpendicular to the tangent at *P* to meet the diameter through *P* in *R*. Prove that *PR* is equal to the latus rectum.

2. *C* is the centre and *A* the vertex of an ellipse. The ordinate and tangent at any point *P* meet the transverse axis in *N* and *T*. Prove that $CN \cdot CT = CA^2$.

Show that any circle passing through *N* and *T* cuts the auxiliary circle orthogonally.

3. If perpendiculars to the asymptotes be drawn from any point on a rectangular hyperbola, prove that their product is constant.

AB and *AC* are fixed straight lines at right angles to one another, and the straight line *BOC* passes through a fixed point *O*, the points *B*, *C* being otherwise undetermined. Prove that the locus of the middle point of *BC* is a rectangular hyperbola, and determine its asymptotes.

4. Form the equations of the perpendiculars of the triangle whose sides are

$$a_1x + b_1y + c_1 = 0,$$
$$a_2x + b_2y + c_2 = 0,$$
$$a_3x + b_3y + c_3 = 0.$$

Deduce that they are concurrent, and show how to find the coordinates of the orthocentre.

5. Investigate the equation to the tangent to the parabola in the form

$$y = mx + \frac{a}{m}.$$

Show that the parabolas

$$y^2 = 4ax, \qquad x^2 = 4by$$

intersect at an angle of 30° if

$$2(a^{\frac{2}{3}} + b^{\frac{2}{3}}) = 3\sqrt{3}a^{\frac{1}{3}}b^{\frac{1}{3}}.$$

6. Find the lengths of the perpendicular from the foci of an ellipse on the tangent at the point whose eccentric angle is ϕ.

Show that they may be obtained by solving the equation

$$u^2 - \frac{2ab}{b'}u + b^2 = 0$$

where a, b are the semi-principal axes, and b' the semi-diameter conjugate to the central vector of the point.

7. Define the hyperbola conjugate to a given hyperbola.

Investigate the equation of the hyperbola conjugate to the hyperbola

$$\frac{x^2}{a^2} - \frac{y^2}{b^2} = 1.$$

8. State the circumstances under which a heavy body will rest on a rough inclined plane without toppling over.

A heavy uniform square lamina balances in a vertical plane on the highest point of a fixed vertical circle in the same plane. The circular rim being rough enough to prevent sliding, show that, if the side of the square be $\frac{\pi r}{\sqrt{3}}$, it can rock through an angle of 60° without falling; r being the radius of the circle.

9. Describe the mechanical power known as the lever. A heavy uniform bar has centre C and one end D. It is placed across two pegs A, B, which are in the same horizontal straight line, in such a manner that they divide the distance CD symmetrically, and it is kept in equilibrium by a weight P suspended from D. If the largest possible value of P is λ^2 times the smallest possible value, show that the distance between the pegs is $\frac{1}{2} \cdot \frac{\lambda - 1}{\lambda + 1}$ of the length of the bar.

10. At a given instant a particle is sliding down a rough plane, inclined at an angle α to the horizon, with a velocity of 32 feet per second; the coefficient of friction being μ, determine the nature of the subsequent motion.

If $\mu = \frac{2}{3}$ and $\alpha = 30°$, for how many seconds will it move before coming to rest?

11. If a ball be projected in vacuo, show that the velocity at any point is that which it would acquire if dropped freely to that point from the directrix to its path.

A ball is projected and a second ball also from the same point, and in the same direction, and with a velocity equal to the vertical velocity of the first ball. Prove that the path of the second passes through the focus of the path of the first.

12. A weight W lbs. is raised from rest through a vertical height H feet in T seconds, when it again comes to rest, by means of a chain which sustains a uniform tension for a part of the time and then becomes slack. Show that the weight ascends with acceleration

$$\frac{2gH}{gT^2-2H} \quad \text{for} \quad \frac{gT^2-2H}{gT} \text{ seconds.}$$

Further show that if the safe tension of the chain be P lbs. weight the time T must not be less than

$$\frac{v}{g}\sqrt{\frac{P}{P-W}} \text{ seconds}$$

where v is the velocity due to the height H.

VII. PART III. SECOND PAPER.

[*Full marks will be given for about* two-thirds *of this paper.*]

1. Show how to draw two tangents to a parabola from a given external point.

Prove that, if three tangents form a triangle PQR, and p is the point of contact of QR, and S the focus, then

$$Qp : Rp = PR \,.\, SQ : SR \,.\, PQ, \text{ the point } p \text{ being between } Q \text{ and } R.$$

2. Prove that, if CP, CD are conjugate semi-diameters of an ellipse whose foci are S and S', then $SP \,.\, S'P = CD^2$.

Prove also that, if the tangents at P and D meet in T, and if SP is produced to R so that $PR = SP$, then the triangles STR, SDT are similar.

3. Define the asymptotes of an hyperbola.

Prove that, if the tangent at P meets one asymptote in T, and the focal radius SP meets the same asymptote in U, the triangle STU is isosceles.

4. Find the angle between the two straight lines whose equations are

$$y = mx + b \quad \text{and} \quad y = m'x + b'.$$

Form the equation of the straight line drawn through the point $x = 1$, $y = 2$ at right angles to the line $x + 2y = 3$.

5. Find the equation of the circle which has a given centre and radius.

Prove that the locus of a point from which the tangents to two given circles are in a constant ratio is a circle.

6. Find the equation of the locus of the middle points of the chords of the parabola $y^2 = 4ax$ which make an angle θ with the axis of x.

Prove that the locus of the middle points of chords of $y^2 = 4ax$, drawn through a given point, is a parabola, and explain the connexion between the two results.

7. Find the equation of the normal at any point (x', y') on the ellipse $x^2/a^2 + y^2/b^2 = 1$.

Prove that the normals at the extremities of the chords $lx/a + my/b = 1$ and $x/la + y/mb = -1$ meet in a point, and find its co-ordinates.

8. Define the moment of a force about a line.

A rigid uniform triangular board is supported by three equal vertical strings whose upper ends are attached to the corners of an equal triangle fixed in a horizontal plane, and a sphere of given weight is placed at a given point on the board. Determine the tensions of the strings.

9. Explain what is meant by the *coefficient of statical friction* between two bodies.

A rigid uniform bar rests against a vertical wall and a horizontal floor, the vertical plane containing the bar being at right angles to the wall. Prove that, if the bar is on the point of slipping down, the angle α which it makes with the horizontal is given by the equation

$$\tan\alpha = \tfrac{1}{2}(1/\mu' - \mu)$$

where μ and μ' are the coefficients of friction at the upper and lower extremities.

10. A smooth sphere of mass m moving with velocity u impinges directly on a smooth sphere of mass m' at rest. Find the velocities of the spheres after impact, the coefficient of restitution being $\tfrac{1}{4}$.

Prove that, if I is the magnitude of the impulse between the spheres, the kinetic energy lost in the impact is $\tfrac{3}{8}\,Iu$.

11. Prove that a particle moving freely under gravity describes a parabola.

If the particle is projected at an elevation α and strikes an inclined plane through the point of projection at right angles, the inclination θ of the plane to the horizontal is given by the equation

$$\tan(\alpha - \theta) = \tfrac{1}{2}\cot\theta.$$

12. Two particles of equal mass are connected by an inelastic string, and held near to each other on a smooth horizontal table. A ring of mass equal to that of either particle is threaded on the string and hangs just over the edge of the table. Prove that, when the particles are let go, the ring descends with acceleration $\tfrac{1}{3}g$.

MATHEMATICAL EXAMINATION PAPERS

FOR ADMISSION INTO

Royal Military Academy, Woolwich,

JUNE, 1899.

I. PART I. FIRST PAPER.

[*Great importance will be attached to accuracy.*]

1. Reduce to their simplest forms: $\frac{1}{2}+\frac{2}{3}+\frac{3}{4}$, $2\frac{3}{8}-\frac{9}{20}$, and $2\frac{2}{7}-\frac{24}{161}$, and divide the product of these three expressions by $\frac{13}{4}+\frac{23}{30}-2$.

2. A tradesman, having promised me $12\frac{1}{2}$ per cent. reduction from a bill for £16. 16*s.* 8*d.* takes off only 10 per cent. By what sum am I entitled to further diminish his bill? And how much per cent. is this of the account rendered?

3. Define the *average* of a set of numbers, and find the average of 213, 217, 199, 201, 208, 209, and 211.

If the height of a number of men is measured; and the average of the first six is 5 ft. 5 in., of the next seven, 5 ft. $5\frac{1}{2}$ in., and of the next eleven, 5 ft. $4\frac{3}{4}$ in.; what is the average height of the whole number of men to the nearest hundredth of an inch?

4. Two trains start at the same time, one from Liverpool to Manchester, and the other in the opposite direction, and running steadily complete the journey in 42 minutes and 56 minutes respectively. How long is it from the moment of starting before they meet?

5. A certain number is the product of 12 and another factor. The number is divisible by 16 and by 17. Show that the second factor is divisible by 17, but not necessarily by 16.

State a general proposition of which this is a particular case.

6. Multiply together any three consecutive whole numbers, all greater than 100. Add to this product the middle number, and show (by extracting the cube root or otherwise) that the result is a perfect cube.

If a is the smallest of the numbers, state and prove the result just given in a general form.

7. Divide $x^7 - 729x$ by $x^2 - 3x + 9$, and express the quotient in its simplest real factors.

8. Solve the equation $\dfrac{x-1}{5} = 8 - \dfrac{x+1}{3}$.

Also if the square of the first side of the above equation is equal to the second side (unaltered), find the *two* values of x; and explain why the first equation has only one solution and the second two.

9. Find the values of x and y in terms of a and b from the equations

$$x - \frac{y}{2} = \tfrac{1}{2}\left(a - \frac{b}{2}\right);$$

$$x + y = \tfrac{1}{3}(a + b - 1).$$

If the values of x and y which are to satisfy this equation are 50 and 51, what must be the values of a and b?

10. An arithmetical and geometrical progression have the same first term 10; and the common difference of the former series, which is 7, is equal to the common ratio of the latter series. Find if any of the numbers 3430, 3455, 3475 belong to either of these series, and, if so, to which.

If $-a$ and $+a$ are two terms in an arithmetic progression, and the number of intermediate terms is $2n$, find the value of the two terms which are numerically smallest.

11. Find the number of permutations of four different things taken all together, giving your *reasons* in full.

A boy, fresh from school, boasts that he is able to distinguish the four brands of champagne (of which he knows the names) in his father's cellar. Accordingly his father fills him a glass of each wine without allowing him to see the labels on the bottles, and requires him to name each variety.

The boy gives the four names and gives each wrong. Determine in how many different ways he can do this, and in how many ways he could have named at least one right.

12. Prove the Binomial Theorem for a positive integral index.

Employ it to show that, to the nearest pound, the interest on £10,000 for 10 years at 2 per cent. compound interest per annum is £2190.

II. PART I. SECOND PAPER.

N.B.—*In questions on Geometry ordinary abbreviations may be employed, but the method of proof must be geometrical. Proofs other than Euclid's must not violate Euclid's sequence of propositions. In the absence of special directions to Candidates, any of the propositions within the limits prescribed for Examination may be used in the solution of problems and riders.*

1. Name and define the different four-sided figures of which Euclid takes note.

If the middle points of every pair of adjacent sides of a rhombus be joined, what is the four-sided figure so formed? Prove that your answer is correct.

2. Through a given point draw a straight line parallel to a given straight line.

Construct a triangle, having given one angle and the lengths of the perpendiculars from the other two angles on the opposite sides.

3. If a straight line be divided into two equal parts, and also into two unequal parts, prove that the rectangle contained by the two unequal parts, together with the square on the line between the points of section, is equal to the square on half the line.

Compare the lengths of the two unequal parts, if the rectangle contained by them is eight times the square on the line between the points of section.

4. Draw a tangent to a circle from an external point.

Given a circle and a straight line, find a point on the line such that, if tangents be drawn to the circle, the chord of contact will subtend a given angle at the circumference. Is it always possible to solve this problem? Give your reasons.

5. A chord AB of a circle is produced to C, so that BC is 7 inches in length. The tangent from C is one foot in length. What is the length of AB?

6. In a given circle place a straight line equal to a given straight line which is not greater than the diameter of the circle.

Show that the straight line subtends an angle of 30° at the circumference if the given straight line be equal to the radius.

7. Find in what ratio the bisector of any angle of a triangle divides the opposite side.

The edge of your paper prevents you from completing *one* angle of a triangle. Show how to draw as much of the internal and external bisectors of that angle as will appear upon your paper, noting carefully what points in your construction you assume to be within the limits of your paper.

8. With a pair of scissors a square, 36 square inches in area, is divided by a straight cut through one angle and the middle point of an opposite side. Find the areas of the two portions.

If another straight cut is made through two opposite angles, so as to cross the first cut, find the areas of the four portions.

9. An equilateral triangle, 10 inches in the side, revolves about one side. Find to the nearest cubic inch the volume of the solid generated (*i.e.*, the volume within the surface traced out by the other two sides).

10. If the area of one face of a regular tetrahedron is A, what is its volume?

III. PART I. THIRD PAPER.

1. Define the *radian*.

The sun is 93 million miles distant, and subtends an angle of 0·00466 of a radian. Find its diameter.

2. Prove the relation $\sin^2\theta + \cos^2\theta = 1$, and show that the fractions $\sin\theta/(1-\cos\theta)$ and $\sin\theta/(1+\cos\theta)$ are reciprocal. (The sign / denotes the fraction-line.)

Hence find a third proportional to $1+\cos\theta$ and $\sin\theta$, when θ approaches indefinitely near to two right angles.

3. Prove, by means of a geometrical figure, that $\sin 2\theta = 2\sin\theta\cos\theta$, 2θ being an acute angle.

If Q is the quotient obtained by dividing $\sin^3\theta + \cos^3\theta$ by $\sin\theta + \cos\theta$, prove that $\sin 2\theta = 2(1-Q)$.

4. Ascertain and write down a general formula giving all angles, both positive and negative, of which the cosine is equal to $\cos A$.

Find the positive values of the angle A, less than 360°, for which $\cos(A+20^\circ) = \cos(3A+60^\circ)$.

5. The cosecant of an acute angle is $\frac{5}{4}$. What are its sine, cosine, and tangent? What changes in the sine, cosine, and tangent occur, if the angle is not acute but obtuse?

Considering acute angles only, prove that the angle

$$\operatorname{cosec}^{-1}\frac{5}{4} - \cot^{-1}4$$

is less than half a right angle.

6. Prove that in any triangle $\cos^2\frac{A}{2} = \frac{s(s-a)}{bc}$.

Find the cosine of half the greatest angle, and the sine of the least angle in a triangle of which the sides are 109, 91, 60.

7. Find an expression for the radius of the inscribed circle of a triangle in terms of the sides.

If the sides are 3, 4, and 5 inches in length, in what ratios do the points of contact divide the sides?

8. Show how to solve a triangle, having given two sides and the angle opposite one of them. Determine the condition that the triangle may exist (i.) when the given angle is acute, (ii.) when it is obtuse.

If $a = 1000$ inches, $b = 353$ inches, $B = 20° 35'$, find, with the aid of the tables supplied, the angles A and C, taking A to be obtuse.

9. Write down the logarithms of the numbers 1234, 2345, 345·1, 45·12, 5·123; and calculate, as accurately as your tables permit, the value of the fraction

$$\frac{1234 \times (2345)^2 \times (345 \cdot 1)^3}{\sqrt[4]{45 \cdot 12} \times \sqrt[5]{5 \cdot 123}}.$$

10. Define the logarithm of a number to the base x.

What is the relation between two numbers, if the sum of their logarithms to base x is (1) zero, (2) unity, (3) three times the lesser of the two logarithms?

IV. PART II. FIRST PAPER.

[*Full marks will be given for about* three-quarters *of this paper.*]

1. Find the condition that the roots of the quadratic equation

$$ax^2 + 2bx + c = 0$$

may be real and distinct.

If α and β are the roots of the above equation, form the equation whose roots are $\alpha + 1/\beta$ and $\beta + 1/\alpha$.

2. Find the number of sets of r things which can be formed with n different things.

Prove that the number of ways in which nine different things can be distributed into sets of three is 280.

3. Prove that the angle subtended at the centre of a circle by any chord is double of the angle in the greater segment cut off by the chord.

If AB is the chord, C the centre of the circle, P any point on the greater segment cut off by AB, and if PA and PB meet the diameter perpendicular to AB in D and E, then the triangles DPE and DCA are similar.

4. On the sides AB and AC of a triangle ABC points F and E are taken so that $CE : EA = AF : FB = 2 : 1$. Find the ratio of the areas of the triangles ABC and AEF.

Also prove that, if L and M are the feet of the perpendiculars let fall on BC from E and F, then $EL : FM = 2 : 1$.

5. The base BC of a triangle ABC is divided at D so that

$$BD : CD = n : 1.$$

Prove that $(n+1) \cot ADC = n \cot BAD - \cot DAC$.

6. A church spire is seen in a direction due North of a station A at an elevation of 17°, and from a station B 120 feet due East of A the spire bears 23° West of North. The two stations and the foot of the spire being at the same level, determine the height of the spire.

7. Prove that a straight line which is perpendicular to each of two straight lines at their point of intersection is perpendicular to the plane containing them.

8. Prove that, if each of the edges of a tetrahedron is perpendicular to the opposite edge, the lines joining the middle points of pairs of opposite edges are equal in length.

9. Define a conic section by means of a focus and directrix.

Prove that the line joining the focus to the point of intersection of any chord with the corresponding directrix is one of the bisectors of the angle between the radii drawn from the focus to the extremities of the chord.

10. Prove that the semi-latus rectum of a conic is a harmonic mean between the segments of any focal chord.

11. Prove that the tangent at P to a conic whose focus is S and the line through S perpendicular to SP meet on the directrix.

If R is the point where they meet, SY perpendicular to the tangent PR, Q a point where SR meets the conic, and QK parallel to PR meets the directrix in K, then $PY : SP = SQ : QK$.

12. Prove that the projection on the focal radius of the normal terminated by the transverse axis is equal to the semi-latus rectum.

Given a point of a conic, the tangent at the point, a focus, and the length of the semi-latus rectum, construct the direction of the transverse axis.

V. PART II. SECOND PAPER.

[*Full marks may be obtained for about* three-quarters *of this paper. Gravitational acceleration may be taken equal to* 32 *feet per second per second.*]

1. State the theorem known as the parallelogram of forces.

A uniform cubical block of wood of weight W rests on a smooth inclined plane with four of its edges parallel to the line of greatest slope, and another of its edges in contact with a smooth vertical wall. If the inclination of the plane be 45°, find the pressure on the wall.

2. Define a couple and its moment.

Show that two couples in the same plane are in equilibrium if their moments are equal and opposite.

3. AB and BC are two uniform beams fixed together at B of length 3·72 and 1·92 inches respectively. AB rests on a horizontal plane, and BC lies in the vertical plane containing AB, the angle ABC being 125°. Find graphically the line of the resultant reaction between AB and the plane. The weights of the beams are proportional to their lengths.

4. Find the centre of mass of two bodies, the mass and centre of mass of each being given.

Three weightless wires form a triangle ABC, which is suspended from A. Masses of 2 lbs. and 1 lb. are fixed to the middle points of AB and AC, and one of 3 lbs. to the corner C. What mass must be placed at B so that in equilibrium the point of trisection of BC next B shall be vertically under A?

5. A balance consists of a perfectly uniform isosceles triangle suspended from the vertex, the scale-pans being hung from the ends of the base. Unequal weights are placed in the pans and balance. They are interchanged, and to effect equilibrium one ounce has to be added to one of them. Show that the one scale-pan is half an ounce heavier than the other.

6. What is meant by the angle of friction?

A rough triangular prism is fixed with one face on a horizontal plane. The other faces are inclined to the plane at angles of 55° and 20°. Masses m_1, m_2 are connected by a thread passing over the upper edge of the prism and lying along lines of greatest slope, and rest on the two inclined faces. If the coefficient of friction on each face is tan 25°, find the ratio $m_1 : m_2$ when the equilibrium is critical.

7. Three points are situated at the corners of an equilateral triangle. They all start with velocity V along the sides in the same direction round the triangle. Find the velocity of any one relative to any other.

8. Give Newton's Second Law of Motion.

On what units does the unit of force depend? The foot, second and pound being the units of length, time and mass, how many units of force are there in the weight of one pound?

What experimental result shows that the weights of bodies are proportional to their masses?

9. A balloon is ascending with uniform acceleration equal to $\frac{1}{8}g$. Find the weight indicated by a spring balance in the balloon-car, on which a mass of one pound is hung.

10. What is a horse-power?

At what horse-power is an engine working which is dragging a load of six tons up an incline of one in ten at a rate of fifteen miles per hour? (Friction and all retarding forces except gravity to be omitted.)

11. A cone of semi-vertical angle 30° is fixed with its axis vertical, and vertex upwards. A bead is attached to the vertex by a thread of length two feet, and rests on the smooth surface of the cone. What velocity must be given to it that it may move round the cone in a horizontal circle without causing any pressure on the surface?

12. Two particles, moving with equal velocities V in perpendicular directions, collide and stick together. Their masses being in the ratio of 4 to 3, find the common velocity after impact.

VI. PART III. FIRST PAPER.

[*Full marks will be given for about* two-thirds *of this paper.*]

1. Prove that if through any point P on an ellipse a line is drawn perpendicular to the major axis to meet the auxiliary circle in p, on the same side of the major axis as P, the tangents at P and p to the ellipse and circle meet in a point on the major axis.

Prove also that if q corresponds to Q in the same way as p to P, and if the tangents to the ellipse at P and Q meet in T and the tangents to the circle at p and q meet in t, then Tt is parallel to the minor axis.

2. Prove that the triangle formed by the two asymptotes of a hyperbola and any tangent is of constant area.

Prove that if LL' and MM' are two tangents which meet one asymptote in L and M, and the other in L' and M', then LM' is parallel to $L'M$.

3. Find the cosine of the angle between the two straight lines whose equations are

$$ax+by+c=0 \quad \text{and} \quad a'x+b'y+c'=0.$$

Form the equation of the two straight lines passing through the point of intersection of the lines $3x+4y+7=0$ and $4x-3y+1=0$ which make angles of 60° with the former line.

4. Find an expression for the area of a triangle in terms of the co-ordinates of its angular points.

Find the area of the triangle whose sides are $y=x$, $y=2x$, $x+y-6=0$.

5. Form the equation of a circle which passes through the origin, has the line $y=mx$ for a diameter, and touches the line $my+x=a$.

6. Find the equation of the normal to the parabola $y^2=4ax$ at any point.

Prove that, if any two chords PQ, PR at right angles to each other are drawn through the point $P(x', y')$ of the parabola, QR meets the normal at P in a point on the line $y+y'=0$.

7. Find the equation of the diameter of the ellipse $x^2/a^2+y^2/b^2=1$ which is conjugate to the diameter $y=mx$.

Prove that the equation of the locus of a point dividing in the ratio 2 : 1 the chords of the ellipse which make an angle θ with the major axis is

$$\left(\frac{\cos^2\theta}{a^2}+\frac{\sin^2\theta}{b^2}\right)\left(\frac{x^2}{a^2}+\frac{y^2}{b^2}-1\right)+8\left(\frac{x\cos\theta}{a^2}+\frac{y\sin\theta}{b^2}\right)^2=0.$$

8. Determine the conditions that a system of forces acting in one plane on a rigid body may keep it in equilibrium.

A rigid uniform rod is supported by two light strings attached to its extremities and to two fixed points at the same level. Prove that if the strings make equal angles with the vertical they must be equal in length.

9. Explain what is meant by the coefficient of friction between two bodies.

A rigid uniform rod is supported with its lower extremity resting against a rough plane of inclination α by a force applied to its upper extremity in the direction of the lines of greatest slope of the plane. Prove that, if the rod makes an angle $\alpha+\beta$ with the vertical and an angle $\beta(<\alpha)$ with the normal to the plane, the coefficient of friction cannot be less than

$$\tfrac{1}{2}\sec\alpha\sec\beta\sin(\alpha-\beta).$$

10. Show how to find the direction and magnitude of the relative velocity of two points which are moving with given velocities.

Two particles move along two perpendicular lines with uniform accelerations f and g, starting together from the point of intersection of the lines with velocities u and v. Prove that if $ug=vf$ their relative velocity at any time is directed along the line joining them.

11. A particle is projected from a given point in a given direction with a given velocity, and moves under gravity; determine the vertex of its path.

Prove that if a particle is just to clear three parallel vertical walls of heights b, c, b above the level of the point of projection and at distances a apart, the direction of projection must make with the horizontal an angle

$$\tan^{-1}\frac{2}{a}\sqrt{\{c(c-b)\}}.$$

12. A wheel is rolling uniformly on a level road and a smooth heavy bead is free to slide along one of the spokes. Find the least velocity of the wheel for which the bead will be always in contact with the rim of the wheel.

VII. PART III. SECOND PAPER.

[*Full marks will be given for about* two-thirds *of this paper.*]

1. Define a parabola and deduce that the square of the ordinate of a point on the curve varies as the abscissa.

A is the vertex of a parabola, *P* any point on the curve, *M* a point on the axis such that *MPA* is a right angle. Show that if *PN* is perpendicular to *AM* then *MN* is equal to the latus rectum.

2. Prove that the sum of the focal distances of a variable point on an ellipse is constant.

Having given the foci and the length of the major axis of an ellipse, show how to find the points on it at which the line joining the foci subtends a given angle. If the eccentricity of the ellipse is $\frac{1}{2}$, within what limits must the angle lie?

3. Find the equation of the straight line through the points whose co-ordinates are $(3, -1)$, $(-\frac{1}{2}, 2)$.

The x co-ordinate of a point is 3, and it lies on the line joining the origin to the point whose co-ordinates are $(5, 7)$. Find its y co-ordinate.

4. Find the perpendicular distance of the point (x, y) from the line whose equation is $x \cos \alpha + y \sin \alpha = p$.

Find the locus of a point which moves so as always to be at a given distance of 10 units from the straight line $4x + 3y = 25$.

5. Find the co-ordinates of the centre, and the radius, of the circle whose equation is $x^2 + y^2 + 2y + 1 = 2x$.

Where are the intersections of the circles

$$x^2 + y^2 = 4x + 6y \quad \text{and} \quad x^2 + y^2 + 2x + 2y = 0?$$

6. Find the equations of the tangents to the parabola $y^2 = 4x$ at the points whose x co-ordinates are each equal to 1.

Show that they are at right angles and intersect where the directrix cuts the axis.

7. Defining the ellipse as the locus of a point the sum of whose distances from two fixed points is constant, find its equation, taking the line joining the fixed points as axis of x, and the middle point of this line as origin.

If $(x_1, y_1)(x_2, y_2)$ are the co-ordinates of two points on the ellipse

$$\frac{x^2}{a^2} + \frac{y^2}{b^2} = 1,$$

show that

$$\frac{(x-x_1)(x-x_2)}{a^2} + \frac{(y-y_1)(y-y_2)}{b^2} = \frac{x^2}{a^2} + \frac{y^2}{b^2} - 1$$

is the equation of the line joining them. Deduce the equation of the tangent at any point.

8. Give an example of (1) stable, (2) unstable, (3) neutral equilibrium.

A number of cubical blocks are piled one on the other so as to form a staircase, the breadth of each step being 2″, and the side of each block being 1 foot. How many can be piled before the whole begins to topple?

9. A board in the shape of an equilateral triangle is hung horizontally by three vertical strings attached to its corners. Show where to place a weight so that the tensions in the strings may be as 1 : 2 : 3. (Neglect the weight of the board.)

10. An elevator-cage is suspended by a rope passing over a smooth pulley and having a counterpoise equal to the weight of the cage at its other end. The rope between the weight and the pulley passes through the cage and can be handled by anyone within. A person of weight W steps inside, and by the friction of his hands on the rope reduces the downward acceleration to $\frac{1}{4}g$. Find the tension on the rope produced by the friction of his hands. Take the weight of the cage to be W, and that of the rope to be negligible.

11. A wedge has one face resting on a smooth table and is free to move. A smooth particle of half its mass is placed on the other face, one foot from the edge. How long will it take to reach the table, the angle of the wedge being 30°?

12. In Atwood's machine where two masses m_1 and m_2 ($m_1 > m_2$) are connected by a light string passing over a smooth pulley, find the tension of the string during the motion.

If the pulley ascends with a uniform acceleration equal to $\frac{1}{8}g$, find the change in the tension.

MATHEMATICAL EXAMINATION PAPERS

FOR ADMISSION INTO

Royal Military Academy, Woolwich,

NOVEMBER, 1899.

I. PART I. FIRST PAPER.

[*Great importance is attached to accuracy.*]

1. From $\frac{1}{5}-\frac{1}{7}+\frac{1}{9}$ take $\frac{1}{3}(\frac{1}{3}-\frac{1}{5}+\frac{1}{7})$, and simplify the fraction

$$\frac{\frac{1}{14}-\frac{1}{2}+\frac{78}{119}-\frac{703}{910}}{\frac{1}{20}+\frac{1}{60}-\frac{1}{28}}.$$

2. If one perch of wire weigh 24 lb. 12 oz. and one mile of the same wire cost £27. 10*s*. 0*d*., how much would 8 cwt. 2 qr. 8 lb. of the wire cost?

3. Find the present worth of £504. 15*s*. 0*d*. due 73 days hence at $4\frac{3}{4}$ per cent.

4. Extract the square root of 108241.

How many whole numbers between 100 and 100,000 are perfect cubes?

5. A certain length of pathway has to be constructed; it is found that three men can construct one-fifth all but one mile in two days, whilst 18 men can construct one mile more than two-fifths in one day. What is the length of the path?

6. Two numbers differ by 6; show that if 9 be added to their product the sum will be a square number. If the square root of this sum be 13, find the two numbers.

7. Multiply $2x^2 - 19x + 35$ by $2x^2 - 13x + 15$, and divide the result by $4x^2 - 16x + 15$.

8. Solve the equations

$$\text{(i.)}\quad \frac{2x-1}{5} - \frac{3x+1}{16} = \frac{5}{2};$$

$$\text{(ii.)}\quad \frac{2x^2-1}{5} - \frac{3x+1}{4} = \frac{9}{10}.$$

9. Solve the equations

$$\left.\begin{aligned} ax + by &= a^2 + 2ab - b^2 \\ bx + ay &= a^2 + b^2 \end{aligned}\right\}.$$

10. Having given the first term and the common difference of an Arithmetic Progression, find the expression for the sum of n terms of the series.

Find the sum of 25 terms of the series

$$1 + 2 + 3 + 4 + \ldots,$$

and find the value of n when the sum of n terms of the series

$$1 + 2 + 4 + 8 + \ldots$$

exceeds by 186 the sum of 25 terms of the preceding series.

11. Write down the expression for the number of permutations of n things taken r together, and hence deduce the expression for the number of combinations of n things taken r together.

Of 15 men, 10 can row and cannot steer, and five can steer and cannot row; find how many boats crews of eight rowers and a coxswain can be formed out of the 15 men.

12. Write down the expression for the coefficient of x^r in the expansion of $(1+x)^n$, and find the greatest coefficient in the expansion of $(1+x)^8$.

Also find the value of x when the fifth term in the expansion of $(1+x)^{\frac{7}{2}}$ is equal to the number 70.

II. PART I. SECOND PAPER.

N.B.—*In questions on Geometry ordinary abbreviations may be employed, but the method of proof must be geometrical. Proofs other than Eucl.d's must not violate Euclid's sequence of propositions. In the absence of special directions to Candidates, any of the propositions within the limits prescribed for Examination may be used in the solution of problems and riders.*

1. Define an *isosceles triangle, a circle,* and *parallel straight lines.*

Construct an isosceles triangle for which each of the equal sides shall be double the base.

2. Show that any two sides of a triangle are together greater than the third side.

Show that of triangles described on a given base, and having a given area, that which has the least perimeter it isosceles.

3. Divide a given straight line into two parts so that the rectangle contained by the whole and one part may be equal to the square on the other part.

If the line be divided so that nine times the rectangle contained by the whole and one part is equal to four times the square on the other part, find the point of division.

4. Prove that the angles contained by a tangent of a circle and a chord of the circle drawn from the point of contact of the tangent are respectively equal to the opposite angles subtended at the circle by the chord.

Describe two circles to touch two given circles, the point of contact with one of these given circles being given.

5. The radii of two intersecting circles are respectively 15 inches and 13 inches, and the common chord of the circles 24 inches long. What length of the line joining their centres lies within both circles?

6. Construct the centre of the circle inscribed in a given triangle.

Prove that, if ABC is the triangle, I the centre of the circle, D the middle point of BC, L the point where AI produced meets BC, and P the foot of the perpendicular from A on BC, then L lies between P and D, the sides AB and AC being unequal.

7. State the relation between the areas of similar triangles and the lengths of their sides.

On the base BC of a triangle ABC as diameter a circle is described cutting the sides AB and AC in P and Q. Prove that, if the triangle APQ is half the triangle ABC, then PQ is the side of a square which can be inscribed in the circle.

8. In one side AB of a parallelogram $ABCD$ a point P is taken so that $3AP = 5PB$. Find the point Q in CD such that the area of the figure $APQD$ is half the area of the parallelogram. Compare the areas of the triangles PQC and PQD with the area of the parallelogram.

9. A frustum of a pyramid is contained between two parallel planes distant 9 feet apart, and the areas of the parallel faces are 5 square feet and 20 square feet. Determine the height of the pyramid and the volume of the frustum.

10. A square whose side is of length 1 foot revolves about one diagonal. Determine, to the nearest square and cubic inch respectively, the area of the surface, and the volume of the figure traced out. ($\pi = 3{\cdot}14159$.)

III. PART I. Third Paper.

1. Define the *sine* and *cosine* of any angle.

Show that the sine of an acute angle is greater or less than the cosine, according as the angle is greater or less than 45°.

2. Prove by geometrical constructions that for all values of A

$$\text{(i.)}\quad \sec^2 A = 1 + \tan^2 A\,;$$
$$\text{(ii.)}\quad \sin A = -\cos(90^\circ + A).$$

3. If D be the number of degrees in the smallest positive angle having its sine $= \frac{4}{5}$, find the values of

$$\cos(180 - D)^\circ,\quad \sin(180 + D)^\circ,\quad \tan(360 - D)^\circ.$$

Also write down a general formula including all angles whose

$$\text{sine} = +\tfrac{4}{5} \text{ and cosine} = -\tfrac{3}{5}.$$

4. Write down (without proof) formulae expressing $\sin(\theta + \phi)$ and $\cos(\theta + \phi)$ in terms of the sines and cosines of θ and ϕ, and thence express $\tan 2\theta$ in terms of $\tan\theta$.

Show that

$$\sin 2\theta = 1 - \cos 2\theta \cot\left(\frac{\pi}{4} + \theta\right).$$

5. If θ be the circular measure of an angle, prove that, as θ is indefinitely diminished, the ratios $\theta : \sin\theta$, $\theta : \tan\theta$, approach to the limit unity.

A man standing beside one milestone on a straight road, observes that the foot of the next milestone is on a level with his eyes, and that its height subtends an angle of $2'\ 55''$. Find the approximate height of that milestone. ($\pi = \frac{22}{7}$.)

6. Write down the values of $\sin 36^\circ$ and $\cos 36^\circ$ as given by your tables. Calculate the sum of the squares of these numbers to six decimal places, and explain why the result differs from unity.

ABCDE is an equilateral and equiangular pentagon. Show that if the distance of A from B or E be 34 inches, its distance from C or D will be 55 inches nearly.

7. In any triangle prove that

$$\frac{a}{\sin A} = \frac{b}{\sin B} = \frac{c}{\sin C}.$$

WOOLWICH ENTRANCE EXAMINATION.

If BC be 25 inches, and CA be 30 inches, and if the angle ABC be twice the angle CAB, find the angles of the triangle ABC, and show that the length of the third side AB is 11 inches.

8. P, Q, R are three villages. P lies 7 miles to the north-east of Q, and Q lies $11\frac{1}{4}$ miles to the north-west of R. Find the distance and bearing of P from R.

9. Define the logarithm of a number to a given base, and show that the logarithm of a number to a base b may be obtained by dividing its logarithm to a base a by $\log_a b$.

Find the logarithm of 83 to base 19 as accurately as your tables permit.

If $x = \log_a bc$, $y = \log_b ca$, and $z = \log_c ab$, prove that

$$xyz = x + y + z + 2.$$

10. Point out the use of logarithms in facilitating numerical calculations, and employ the tables to obtain the values of

(i.) $\dfrac{327{\cdot}4 \times \sqrt[3]{0{\cdot}0006}}{\sqrt{62{\cdot}81}}$;

(ii.) $\dfrac{\sin 25^\circ \cos 37^\circ}{\tan 130^\circ}$.

IV. PART II. FIRST PAPER.

[*Full marks will be given for about* three-quarters *of this paper.*]

1. Investigate the connexion between the coefficients of the equation

$$ax^4 - bx^3 + cx^2 - dx + e = 0$$

and the simple symmetric functions of its roots.

Express (1) the sum of the squares, (2) the sum of the fourth powers of the roots of the equation in terms of the coefficients.

2. Find the general term of the series

$$(x+3)^{n-1} + (x+3)^{n-2}(x+2) + (x+3)^{n-3}(x+2)^2 + \dots$$
$$+ (x+3)(x+2)^{n-2} + (x+2)^{n-1}$$

when it has been arranged in descending powers of x.

3. Prove that two circles cannot intersect in more than two points.

Draw a circle to pass through two given points, and to touch a given circle.

How must the positions of the points be limited?

4. A triangular plot of grass has sides a, b, c feet in length; it is bordered by a pathway m feet in breadth. Find the area of the pathway, and if the corners be rounded off by arcs of circles so that no part of the boundary is more than m feet from the grass, find by how much the area will be diminished.

5. Find an expression for $\cos(\alpha+\beta+\gamma)$ in terms of sines and cosines of α, β and γ.

Prove the identity

$$\cos\alpha\cos(\beta+\gamma) + \cos\beta\cos(\gamma+\alpha) + \cos\gamma\cos(\alpha+\beta)$$
$$= \cos(\alpha+\beta+\gamma) + 2\cos\alpha\cos\beta\cos\gamma.$$

6. A man has before him on a level plane, a conical hill of vertical angle 90°. Stationing himself at some distance from its foot he observes the angle of elevation α of an object which he knows to be half way up to the summit. Show that the part of the hill above the object subtends at his eye an angle

$$\tan^{-1}\frac{\tan\alpha(1-\tan\alpha)}{1+\tan\alpha(1+2\tan\alpha)}.$$

7. Prove that there can be but one perpendicular to a plane from a point without the plane.

8. The base of a pyramid is an equilateral triangle, and the angles at the vertex are right angles. Show that the sum of the distances of the faces from any point in the base is constant.

If a side of the base be one foot, calculate the volume of the pyramid to the nearest cubic inch.

9. The tangents at points P, Q of a conic intersect in T. Prove that TP, TQ subtend equal or supplementary angles at a focus, and separate the cases.

10. Prove that all chords of a conic drawn through any point are cut harmonically by that point and its polar with respect to the conic.

A is the vertex, and AB the major axis of a conic. Let the tangent at any point P of the curve meet BA produced in T. Join PA, PB, and produce these lines to meet a straight line drawn through T perpendicular to AB in R and Q. Prove that $TR = TQ$.

11. P is a point on a conic whose vertex is A and focus S, and G is the foot of the normal through P. Prove that $SG = eSP$.

12. The ratio of the rectangles contained by the segments of any two intersecting chords of a conic is equal to that of the rectangles contained by the segments of any other two chords parallel to the former, each to each.

A, B, C, D are four points on a conic, which lie upon a circle. Show that the lines AB, CD are equally inclined to the axis, and that the same property is enjoyed by two other pairs of lines.

V. PART II. Second Paper.

[*Full marks may be obtained for about* three-quarters *of this paper. Gravitational acceleration may be taken equal to* 32 *feet per second.*]

1. State the triangle of forces.

Three forces act along the sides AB, BC, and AC of a triangle ABC, and are represented in magnitude by AB, BC, and twice AC respectively. Find the magnitude and line of action of their resultant.

2. If three coplanar forces are in equilibrium, prove that they must either be parallel or all meet in one point.

A uniform heavy rod rests with its extremities on two smooth inclined planes sloping towards each other at inclinations to the horizon of $30°$ and $60°$. Find the inclination of the rod to the horizon in the position of equilibrium.

3. The uniform rod AB, 4 feet long, is hinged to a vertical wall at A, and supported in a horizontal position by the string CD attached to the rod at C and the wall at D, a point 4 feet above A. If the weight of the rod is 10 lb., and AC is 3 feet, find graphically the tension in the string, and the magnitude and line of action of the reaction at the hinge.

4. Find the magnitude and position of the resultant of parallel forces of 1, 2, 3, 4, 5 and 6 lbs. acting in vertical lines at distances 1 foot apart in the same vertical plane, the forces of 1, 3 and 5 lbs. acting upwards, and the forces of 2, 4 and 6 lbs. acting downwards. What is the distance of the resultant from the force of 1 lb.?

5. A wire is bent into the form of a triangle. Give a construction for its centre of mass.

If the sides of the triangle are 5, 12 and 13 inches in length, find the distances of its centre of mass from the two shorter sides.

6. Draw two different systems of weightless pulleys each of which gives a mechanical advantage of 8.

A number of pulleys of equal weight w are arranged in the system in which each pulley hangs by a separate string, and the top string passes over a fixed pulley to allow of the "power" being applied downwards. Prove that the weights of the pulleys may be counterpoised by attaching a weight w to the lowest pulley, and a weight w to the string intended to support the "power."

7. Define *acceleration*, and explain how it is measured.

Find the initial velocity and the acceleration of a body which describes 30 centimetres in the first five seconds, and 50 centimetres in the next three seconds of its motion, the acceleration being uniform.

8. A body is projected vertically upwards with a velocity of 80 feet per second. Find and represent in a figure its positions after 1, 2, 3, 4 and 5 seconds respectively, and when at its greatest height.

If a second body is projected at the same instant with a velocity of 90 feet per second, how far apart will the two bodies be after 4 seconds?

9. State Newton's Second Law of Motion, and apply it to the following problems:

(i.) A body weighing 80 grains sliding down an incline of 30° from rest describes 5 feet in the first second. Find the retarding force up the plane.

(ii.) A body of weight W is acted on by gravity and by a force $\frac{1}{2}\sqrt{3}W$ acting in a direction inclined at 30° to the upward drawn vertical, find the direction in which the body begins to move.

10. A mass P on a smooth horizontal table is connected with unequal masses Q and R by strings passing over smooth pulleys at opposite edges of the table. Find the acceleration.

If $Q=7$ lb. and $R=2$ lb., the system starting from rest describes 5 feet in the first second. If $Q=4$ lb. and $R=1$ lb., the system describes 4 feet in the first second. From these data determine P and the acceleration of gravity.

11. A body is projected with velocity V in a direction inclined at an angle α to the horizon. Find the time of flight and range on a horizontal plane. Also prove that the former is twice the time taken to reach the greatest height, and the latter is twice the horizontal distance described when at the greatest height.

12. A particle is revolving with speed v in a circle of radius r. State and prove the formulae for the acceleration, pointing out in what direction this acceleration acts.

If the moon revolves about the earth in a circle whose radius is 240,000 miles, performing one revolution in 30 days, find its velocity and acceleration in foot-second units.

VI. PART III. First Paper.

[*Full marks will be given for about* two-thirds *of this paper.*]

1. Prove that the principal ordinate of any point on the parabola is a mean proportional to the abscissa and the latus-rectum.

A diameter PM is drawn through the point P of a parabola so as to meet the tangent at the vertex A in M; MN is drawn at right angles to AP to meet the axis in N. Prove that AN is equal to the latus-rectum.

2. Given one focus and two points of an ellipse, find the locus of the other focus.

If the length of the major axis of the ellipse be given instead of one of the two points, show that the centre lies on a certain circle.

3. Construct the circle whose rectangular equation is

$$x^2+(y-c)^2=c^2.$$

Take a point P on the positive part of the axis of x; bisect OP, O being the origin, in Q and draw a straight line QAB passing through the centre of the circle and cutting it in the points A, B. Find the equation of the circle passing through the points P, A, B; find also its radius, and show that it touches the axis of x at P.

4. Describe the circle

$$x^2+y^2=c^2,$$

and draw the lines

$$(x\sqrt{3}+y-c)(-x\sqrt{3}+y-c)=0.$$

Find the equation of the line joining the feet of the perpendiculars upon these lines from any point (a, b) on the circle.

Prove that this line also passes through the foot of the perpendicular from the point (a, b) upon the line $2y+c=0$.

5. Establish the equation of the normal to a parabola in the form

$$y - mx + 2am + am^3 = 0.$$

Find the co-ordinates of the point at which this straight line is the normal. If y_1 be the ordinate of this point and y_2 that of the middle point of the normal chord through the point, prove that

$$y_1 y_2 + 4a^2 = 0.$$

6. Chords of a parabola are drawn through a given point in its plane. Find the locus of their points of bisection, and discuss the nature, position, and dimensions of the locus for various positions of the given point.

7. Give a geometrical interpretation of the circumstance that the equation of a hyperbola results from the elimination of λ between the two equations

$$ay = \lambda b(x + a),$$

$$ay = \frac{b}{\lambda}(x - a).$$

8. A heavy particle hangs by a thread which is then drawn out of the vertical until it makes an angle with it. The particle is held in this position by a horizontally applied force, which after a time is suddenly withdrawn. If the tensions of the thread before the withdrawal and immediately afterwards be 4 poundsweight and 2 poundsweight respectively, find the weight of the particle.

9. Give an example of equilibrium of a body in which friction is one of the forces, but the limiting friction is not brought into play.

A uniform beam AB of weight w rests with one end A upon a rough horizontal plane and has its other end B attached to a string, which passes over a smooth pulley E and supports a weight P. Find the position of equilibrium for any value of the friction which does not exceed the limiting friction.

10. State Hooke's law in regard to the extension of an elastic cord.

A light extensible cord is attached to two points in a horizontal plane, its unstretched length being equal to the distance between the points. At its middle point is placed a heavy bead so that in the position of equilibrium the parts of the cord each make a very small angle, whose circular measure is α, with the horizontal. Show that if the extension of the cord be proportional to the tension and a force of P pounds be necessary to double the length of the cord, the weight of the bead is approximately $\alpha^3 P$ pounds.

11. In the case of the path of a projectile in vacuo, show that, for a given velocity of projection, there are in general two directions of projection which will cause a given point to be struck. If the two directions coincide, prove that the angle of elevation is $\tan^{-1}\frac{2h}{a}$, where a is the horizontal distance of the point and h the height due to the initial velocity, and show further that the point is situated upon the parabola

$$x^2 = 4h(h - y),$$

the axes being through the point of projection, that of x horizontal and that of y vertical.

12. A uniform chain hangs by one end and sustains a load at the other. Show how to draw a diagram to represent the work done when the chain is wound up so as to raise the load.

VII. PART III. SECOND PAPER.

[*Full marks will be given for about* two-thirds *of this paper.*]

1. Prove that the semi-latus rectum of any conic is a harmonic mean between the segments of any focal chord.

Prove also that the length of any focal chord of a parabola is equal to $4SP$, S being the focus, and P the extremity of the diameter which bisects the chord.

2. Give a construction for drawing two tangents to an ellipse from a given point.

Defining an *asymptote* of a hyperbola as a tangent whose point of contact is infinitely distant, show that the lines joining the centre of a hyperbola to the points where a directrix cuts the auxiliary circle are asymptotes according to this definition.

3. Interpret the constants in the equation of a straight line expressed

(i.) in the form $y = mx + b$;
(ii.) in the form $x = ny + a$.

If the axis of y is measured vertically upwards, prove that $y' - mx' - b$ represents the vertical height of the point (x', y') above the line whose equation is $y = mx + b$. Deduce the length of the perpendicular from the point (x', y') on the line $y = mx + b$.

4. Show that the equation

$$ax^2 + 2hxy + by^2 + 2gx + 2fy + \frac{af^2 + bg^2 - 2fgh}{ab - h^2} = 0$$

represents two straight lines. If these are inclined to the axis of x at angles θ_1 and θ_2, express $\tan(\theta_2 - \theta_1)$ and $\tan(\theta_2 + \theta_1)$ in terms of a, h, b.

5. Find the co-ordinates of the middle point of the chord which the circle $x^2 + y^2 - 2x + 2y = 2$ cuts off on the line $y = x - 1$.

Find also the locus of the middle points of all chords of the circle which are parallel to the line $y = x - 1$.

6. Prove that the equations

$$y^2 = 4a(x + a) \quad \text{and} \quad y^2 = -4b(x - b)$$

represent a pair of parabolas having the same focus and axis. Find the co-ordinates of their points of intersection, and prove that at each of these points the tangents to the two parabolas are at right angles.

7. Define the *eccentric angle* of any point on an ellipse. Find the equation of the chord joining the points on the ellipse

$$\frac{x^2}{a^2}+\frac{y^2}{b^2}=1,$$

whose eccentric angles are θ, ϕ, and deduce the equation of the tangent at the point whose eccentric angle is θ.

Prove that the equation

$$\frac{x^2}{a^2\cos^2\alpha}+\frac{y^2}{b^2\sin^2\alpha}=1$$

represents, for different values of α, a series of ellipses all inscribed in a certain rhombus.

8. A uniform ladder of length l and weight w is being gradually tilted off the ground by a man who can reach up to a height h. If the ground be perfectly smooth where the lower end touches it, and if $l>2h$, prove that the ladder cannot be raised into a vertical position unless its lower end be weighted, and find the least weight that will suffice. If the ground be rough, show by a diagram that as soon as the inclination of the ladder to the vertical is less than the angle of friction the ladder may be supported at a point below its middle point, provided that a force of suitable magnitude and direction be applied to the point of support, and show how to determine this force when equilibrium is limiting, having given its point of application.

9. State Newton's Third Law of Motion, and show how it affords a means of comparing the masses of different bodies.

A body moving with velocity 5 feet per second overtakes another body moving with velocity 1 foot per second. The first body rebounds with velocity 1 foot per second, its direction of motion being reversed, while the second body has its velocity increased to 2 feet per second. Compare the masses of the bodies, and find the coefficient of restitution.

10. A stone is being whirled in a circle in a vertical plane at one end of a light string of which the other end is fixed. When the stone is at the highest point of the circle, the tension in the string just vanishes. Find the tension in the string when the stone is at the lowest point of the circle and when it is at the end of a horizontal diameter, expressing the result in terms of the weight of the stone. If the string be cut at the instant when the stone is ascending vertically, determine in terms of the radius of the circle the height to which the stone will rise.

11. Prove that the work done in raising a number of weights through different heights is equal to the work which would be done in raising a single weight, equal to the sum of the weights, through a height equal to the height through which the centre of gravity of the weights is raised.

A lake whose superficial area is half a square mile has its surface at a height of 95 feet above the sea level. If the water is run off till the surface is lowered by ten feet, calculate the value of the work which can be done in driving machinery by this water in its descent to the sea, a cubic foot of water being assumed to weigh $62\frac{1}{2}$ lb., and work being valued at a half-penny per horse-power per hour.

12. A number of particles are placed on a rectangular sheet of paper. Given the masses of the particles and their distances from two adjacent edges of the sheet, find the distances of their centre of gravity from these edges.

A series of cubes of the same material are piled one above the other on a horizontal plane. The lengths of the sides of the cubes, beginning with the lowest, are $2a$, $2ar$, $2ar^2$, $2ar^3$, and so on in geometrical progression. If there are n cubes in the pile, find the height of the centre of gravity of the whole system above the horizontal plane. If the number of cubes is infinite, r being less than unity, prove that the height of the centre of gravity of the pile is

$$a\frac{1+r^3}{1-r^4}.$$

MATHEMATICAL EXAMINATION PAPERS

FOR ADMISSION INTO

Royal Military Academy, Woolwich,

JUNE, 1900.

OBLIGATORY EXAMINATION.

I. PART I. FIRST PAPER.

[*Great importance will be attached to accuracy.*]

1. Express as single decimals

(1) $0 \cdot 125 \times 0 \cdot 9152 \div 0 \cdot 0715$.

(2) $\left\{ \frac{1}{13} \times 1\frac{5}{7} + \frac{2}{49} \times 1\frac{2}{3} \right\}$ of $\frac{2\frac{2}{3}}{36\frac{6}{7}}$.

2. What income will a man receive from investing £1000 in a $2\frac{1}{4}$ per cent. stock at $112\frac{1}{2}$?

If he sells out his stock at 105, what loss of capital will he sustain?

3. Find the true discount on £3570. 18*s*. 4*d*., due 50 days hence, at $3\frac{3}{4}$ per cent. per annum.

4. A beam, having a square section, is 9 ft. long, and weighs $3\frac{1}{2}$ cwt. A cubic foot of the substance of the beam weighs 32 lbs. What is the thickness of the beam?

5. A dishonest tradesman marks his goods at an advance of 5 per cent. on the cost price, but uses a fraudulent balance, whose beam is horizontal when the weight in one scale is one-fifteenth more than the weight in the other. What is his actual gain per cent.?

6. A man walks a third as far again as a boy in a given time. His stride is 5 centimetres longer than the boy's, and he takes 125 strides per minute, while the boy takes only 100. At what rate in kilometres per hour does each walk?

7. Multiply $7x^2 - 23x + 6$ by $6x^2 - 35x + 11$, and divide the result by $3x^2 - 10x + 3$.

8. Solve the equations

$$\text{(i.) } \frac{x-5}{2} - \frac{2x-11}{6} = \frac{11x-7}{140}; \quad \text{(ii.) } 9x^2 + 17x - 2 = 0.$$

9. Solve the equations

$$c^2x + 2a^2y = (c+a)(cx+2ay) = (c-a)^2.$$

10. Having given the first term and the constant ratio of a geometric progression, find the expression for the sum of the first n terms.

Find the sum of ten terms of the geometric series

$$3 + 6 + 12 + 24 + \ldots$$

What is meant by saying that the sum to infinity of the series

$$1 + \tfrac{1}{2} + \tfrac{1}{4} + \tfrac{1}{8} + \ldots$$

is 2?

11. Find the number of combinations of n things, r at a time, not assuming the formula for the number of permutations.

In how many ways could a party of five scouts be selected from 20 available men? In how many ways could the twenty men be formed into four parties of five scouts, to proceed in different directions?

12. Prove the binomial expansion, n being a positive integer.

If $$(1+x)^n = 1 + n_1x + n_2x^2 + n_3x^3 + \ldots,$$
where $$n_r = n(n-1)\ldots(n-r+1) \div r!,$$
find the value, when $n = 9$, of $1 - n_2 + n_4 - n_6 + n_8$.

II. PART I. Second Paper.

N.B.—*In questions on Geometry ordinary abbreviations may be employed, but the method of proof must be geometrical. Proofs other than Euclid's must not violate Euclid's sequence of propositions. In the absence of special directions to Candidates, any of the propositions within the limits prescribed for Examination may be used in the solution of problems and riders.*

1. Prove that two triangles standing on the same base and between the same parallels are equal in area.

Hence show that if two triangles have two sides of the one equal to two sides of the other and the contained angles supplemental, they are equal in area.

2. Any point, *P*, is taken on a diagonal of a parallelogram, and through the point are drawn two parallels to the sides; prove that the complements of the parallelograms about the diagonal are equal in area.

Ascertain for what position of *P* on the diagonal the area of each of these complements is greatest.

3. Show how to find the centre of a given circle.

Being given an arc of a circle, show how to describe the whole circle.

4. Prove that the angle subtended at the centre of a circle by any arc is double the angle subtended by the arc at any point on the circumference.

From this show that the sum of a pair of opposite angles of a quadrilateral in a circle is two right angles.

5. Show how to divide a given right line into three equal parts.

From the process employed (or otherwise) show that the bisectors of any two sides of a triangle drawn from the opposite vertices trisect each other.

6. Inscribe a circle in a given triangle.

ABC is a right-angled triangle, *A* being the right angle. Prove that the hypotenuse *BC* is equal to the difference between the radius of the inscribed circle of the triangle and the radius of the circle which touches *BC* and the other two sides produced.

7. Prove that the sides about the equal angles of triangles which are equiangular to one another are proportionate, and that those which are opposite to the equal angles are proportionals.

Find the point P in the side AC of a triangle ABC such that the rectangle contained by AP and AC is equal to the square on AB.

8. AB is a chord of a circle, centre C, equal in length to the radius of the circle. If the length is eighty-five inches, find, to the nearest inch, the length of the arc AB.

$$[\pi = 3{\cdot}1416.]$$

9. The cross-section of a wedge, in the form of a right triangular prism, is an equilateral triangle, the side of which is eleven inches; if the length of the wedge is thirty-seven inches, find, to the nearest integer, the number of square inches in the total surface of the wedge.

10. A railway cutting 8 metres deep has to be made, with one side vertical and the other inclined at 30 degrees to the vertical; the bottom is to be 9·4 metres broad. How many cubic metres (to the nearest integer) of earth must be removed per kilometre?

III. PART I. Third Paper.

1. Explain the sexagesimal measure of angles.

The earth revolves once in 24 hours; express in sexagesimal measure the angles through which it revolves (1) in one hour, (2) in one minute of time, (3) in one second of time.

2. Define the *cosine* of an angle, explaining how its algebraic sign is determined. In which quadrants is the cosine positive?

Construct an angle whose cosine is $-\frac{5}{13}$, and find its remaining trigonometric ratios.

3. State and prove the values of the sine, cosine, and tangent of 30°.

A man 6 feet high observes the elevation of a tower to be 30° when he is standing 80 feet from the base of the tower; find (to the nearest foot) the height of the tower.

4. Prove that $\sin^2\theta + \tan^2\theta = \sec^2\theta - \cos^2\theta$, and solve the equation

$$\sec\theta + \tan\theta = \sqrt{3}.$$

5. If
$$x = X\frac{\sin(\theta - \alpha)}{\sin\theta} + Y\frac{\sin(\theta - \phi - \alpha)}{\sin\theta},$$
and
$$y = X\frac{\sin\alpha}{\sin\theta} + Y\frac{\sin(\phi + \alpha)}{\sin\theta},$$
express $x^2 + 2xy\cos\theta + y^2$ in terms of X, Y, and $\cos\phi$.

6. Assuming expressions for the expansions of $\sin(A+B)$ and $\cos(A+B)$ establish the formulae

$$\sin 2A = 2\sin A\cos A$$
$$\cos A - \cos B = -2\sin\frac{A+B}{2}\sin\frac{A-B}{2}.$$

7. One side of a right-angled triangle is 6·432 feet long, and the angle opposite to it is 37° 27′. Find (i.) the area of the triangle; (ii.) the length of the perpendicular from the right angle on the hypotenuse.

8. In any triangle show that

$$4bc \cos^2 \tfrac{1}{2} A = (a+b+c)(-a+b+c).$$

If $a = 17$, $b = 18$, $c = 19$, calculate A as accurately as your tables permit.

9. Prove that the logarithm of a product is the sum of the logarithms of its factors, the base remaining the same.

If $e = 2{\cdot}71828$, employ the tables to find $\log_{10} 2$, $\log_e 10$, and $\log_e 2$.

10. (i.) Solve the equation

$$2x^2 = 16^{x-1}.$$

(ii.) Find the number of digits in 19^{33}.

(iii.) Find the number of zeros following the decimal place in the value of $(\frac{1}{19})^{33}$.

IV. PART II. First Paper.

[*Full marks will be given for about* three-quarters *of this paper.*]

1. Obtain the common difference, the nth term and the sum of n terms of the arithmetic progression whose first and second terms are u and v respectively.

Show that this sum may be written in the form $nu - \frac{n(n-1)}{1.2}\frac{u^2}{s}$ where (if $\frac{v}{u}$ is a proper fraction) s is the "sum to infinity" of the geometric progression $u, v, \ldots$

2. Write down the rth term of the expansion of $(1+x)^n$ in a series of ascending powers of x. Hence obtain, and reduce to its simplest form, the rth term (u_r) in the expansion of $(1-qx)^{-p/q}$.

If p, q are positive integers, p being less than q, and if qx is a positive proper fraction, show that the ratio $u_{r+1} : u_r$ is positive and less than qx. Hence prove that the sum of the terms which follow the rth term ("the remainder after r terms") is less than $\frac{qx}{1-qx}u_r$.

3. Prove that the straight line joining the middle points of two sides of a triangle is parallel to, and is equal to half the length of the third side.

If $ABCD$ be any four-sided figure the three straight lines joining the middle points of AB and CD, of AC and BD, of AD and BC respectively, meet in one point which bisects each of the joining lines.

4. Construct an isosceles triangle having each of the angles at the base double the vertical angle.

A is the vertex and BD the base of the isoceles triangle; C is the point on AB used in the construction; M is the middle point of BD, and AM intersects CD in N. Show that the ratio $AM : NM$ is equal to the ratio of the perimeter of ABD to its base.

5. A, B, C are three angles whose sum is $2S$, and θ is an angle defined by the equation

$$\cos\theta = \frac{\cos A + \cos B \cos C}{\sin B \sin C};$$

prove that

$$\cos \tfrac{1}{2}\theta = \sqrt{\frac{\cos(S-B)\cos(S-C)}{\sin B \sin C}}.$$

6. The lengths of the sides of a triangle are 375 links, 452 links, and 547 links. Find the length of the perpendicular upon the shortest side from the opposite corner, and the radius of the inscribed circle.

7. If a straight line be at right angles to two other straight lines which cut one another then it is at right angles to the plane containing these lines.

8. A solid cube (edge, 10 inches long) is perforated so that a square pyramid just fits into the hole. The pyramid has a base equal to a face of the cube and a height of 20 inches, and when it is fitted into the hole its base lies flush with the surface of the cube, and its vertex projects beyond the face opposite to the base. What is the area of the whole surface (inside and outside) of the perforated cube?

9. State the focus-directrix definition of a conic section. Show how points upon the conic may be found directly from the definition.

An arc, AB, of a circle is bisected by a diameter of the circle. If this diameter be taken as directrix and the point A be taken as focus, find the eccentricity so that the conic may intersect the circular arc in one of the two points that divide it into three equal parts.

10. Given a focus, the corresponding directrix and one point, P, of a conic construct the tangent at P.

Prove the proposition you use for this purpose.

11. PT is a straight line touching a conic at P. TK is at right angles to the focal distance SKP and TH is at right angles to the directrix XH. Prove that $SK : TH$ is equal to the eccentricity.

12. Prove that the semi-latus rectum is the harmonic mean between the two segments of any focal chord.

V. PART II. Second Paper.

[*Full marks will be given for about* three-quarters *of this paper.*]

1. Show how to resolve a given force into two components, one of which is given in magnitude, direction and sense.

A force R is resolved into two components, one of them being in a line making an angle θ with the line of action of R; find the least possible magnitude of the other.

2. A is a point 4 feet to the left of and 1·2 feet vertically above B. A string, attached to the point A, passes over a smooth fixed peg B, and supports a weight of 19 lbs.; at a point C of the string a second string is knotted to the first and pulled downwards with a tension P. Determine the magnitude of P. ($AC = 3{\cdot}2$ feet, $BC = 2{\cdot}7$ feet.)

3. A light rod AB, length 3·2 feet, can turn in a vertical plane about a hinge at A, and supports a weight of 10 lbs. at D, its middle point, and the system is supported in a horizontal position by a light rod CB, C being fixed 2·4 feet vertically under A; find graphically the thrust in the rod CB.

4. Prove that, if the lines of action of two forces intersect, the sum of their moments about any point in their plane is equal to the moment of their resultant about the same point.

A uniform circular disc of weight W is supported in a vertical plane on two smooth pegs, which are so placed that the line joining them subtends a right angle at the centre of the disc, and makes an angle α ($< 45°$) with the horizontal. Determine the pressure on each peg.

5. Find the magnitude and line of action of the resultant of two parallel forces acting in the same sense on a rigid body.

The bar of a common steelyard is 2 feet long, weighs 3 lbs., and is suspended at a point distant 2 inches from one end, and its centre of gravity is distant 3 inches from the same end. Find the greatest weight that can be weighed with this steelyard, the movable weight being 5 lbs.

6. Three pieces of wire of equal weights and unequal lengths are joined together so as to form a triangle. Find its centre of gravity.

Three pieces of wire of equal lengths, weighing 1, 2, and 3 grains respectively, are joined together so as to form a triangle which is hung up by the corner opposite to the heaviest side. Find the depth of the centre of gravity below the corner.

7. Prove the formula for uniformly accelerated motion $s = ut + \frac{1}{2}ft^2$.

A particle moves from a point A in a straight line with uniform velocity v; and a second particle starts from rest at A at the same instant, and moves in the same direction along the same line with uniform acceleration f. Find the greatest distance between the particles before the second overtakes the first.

8. The velocities of two moving points being given, determine their relative velocity.

Two particles move with constant velocities along two straight lines which cut at right angles; prove that, if the second moves through the point of intersection with velocity v at an instant when the first is distant a from that point, the angular velocity of the line joining them when the distance between them is r is av/r^2.

9. A body of mass m slides down a plane of inclination a to the horizontal under gravity and a constant resistance F, acting up the line of slope; find its velocity after descending a height h.

If the body is projected up the plane with this velocity, and moves under gravity and the resistance F, acting down the line of slope; find how far it ascends before coming to rest.

10. Two bodies whose masses are m and m' are connected by a cord passing over a smooth pulley. The pulley is at a height $2h$ above a table, on which is another body of mass $m - m'$, vertically under m', and attached to m' by another cord of length h. The system is let go from a position in which m' is at a height $\frac{1}{2}h$ above the table, and the table does not interfere with the motion of m. Find the velocity of m' just before the second cord becomes tight.

Find also the velocity of m just after the second cord becomes tight.

11. A particle is projected from a given point with a given velocity. Find the range on an inclined plane through the point of projection.

Prove that, if the angle of elevation at starting is such as to make the range as great as possible, the velocity of the particle when it strikes the plane will be as small as possible, the further extremity of the range being at a higher level than the point of projection.

12. A particle of mass m is describing a circle of radius r with uniform speed v. State (without proof) the magnitude and direction of the force acting upon it.

To a string, of which one end is fixed, are attached two particles of equal masses at the further end and the middle point; and the two particles describe circles uniformly, with the two portions of the string always in a straight line. Prove that the tensions of the two portions are in the ratio 2 : 3.

VOLUNTARY MATHEMATICS.

VI. PART III. FIRST PAPER.

[*Full marks will be given for about* two-thirds *of this paper.*]

1. The tangents at two points P, Q on a parabola, whose focus is S, meet at T; show that the angles TSP, TSQ are equal.

If the angle PTQ be one-third of two right angles and the finite line ST cut the parabola at R then the tangents at P, Q, R form an equilateral triangle.

2. Prove that the feet Y, Y' of the perpendiculars SY, $S'Y'$ drawn from the foci S, S' of an ellipse to the tangent at any point P lie on the major auxiliary circle.

From Y another tangent YQ is drawn to the ellipse; show that SQ is parallel to $S'P$.

3. Prove that the area of the triangle formed by the origin and the two points whose rectangular co-ordinates are (x_1, y_1), (x_2, y_2) is $\frac{1}{2}(x_1y_2 - x_2y_1)$.

Find the length of the perpendicular drawn from the origin to the line joining the points whose rectangular co-ordinates are (4, 1), (7, 5).

4. Show that each of the circles

$$x^2+y^2-2x\sqrt{3}+2y=0 \text{ and } x^2+y^2+\frac{2}{\sqrt{3}}x-2y=0$$

circumscribes an equilateral triangle which has one side lying along an axis of co-ordinates.

Find the co-ordinates of the second point of intersection of these circles.

5. Show that the parabolæ whose equations in rectangular co-ordinates are $y^2-ax=0$, $x^2-by=0$ have only one common chord which cuts the curves in real points.

If P and P' be the points on $y^2 - ax = 0$ and $x^2 - by = 0$ the tangents at which are parallel to this common chord, and S, S' be the foci of the parabolæ, then SP is perpendicular to $S'P'$.

6. If S and S' are the foci, SL the semi-latus rectum of an ellipse, and LS' produced cuts the ellipse at X, show that the length of the ordinate of X is $\frac{(1-e^2)^2}{1+3e^2}a$ where $2a$ is the length of the major axis and e is the eccentricity of the ellipse.

7. Find the equation of the tangent to the hyperbola $\frac{x^2}{a^2} - \frac{y^2}{b^2} = 1$ at the point (x', y').

If y_1, y_2 be the ordinates of the points in which this tangent cuts the major auxiliary circle show that $\frac{1}{y_1} + \frac{1}{y_2} = \frac{2}{y'}$.

8. A smooth pencil, whose cross-section is a circle of radius r, is laid upon a flat and smooth ruler whose breadth is $2c$, so that its axis is parallel to the edge of the ruler; the two are now bound together by a string which, passing under the ruler and over the pencil, lies in a plane perpendicular to the axis of the pencil; show that the pressure between the pencil and the ruler is $\frac{4rc}{r^2+c^2}T$ where T is the tension of the string.

9. Enunciate the laws of statical friction.

A pair of step-ladders stands on a rough horizontal plane with each of its ladders on the point of slipping; show that the feet of the ladders are at a distance $\frac{4\mu l}{\sqrt{1+4\mu^2}}$ apart, where μ is the coefficient of friction, and l is the length of each ladder. (The centre of gravity of each of the ladders is to be supposed to be half way up the ladder.)

10. Distinguish between stable, neutral, and unstable equilibrium.

A uniform beam AB, whose weight is W, is moveable in a vertical plane about a hinge at A; to the end B, one end of a string which carries a weight W at its other end is attached; the string is passed over a pulley at a height h ($h > AB$) vertically over A; determine whether the beam is in stable or unstable equilibrium when it is in a vertical position above A.

11. Explain the meaning of the terms Work, Kinetic Energy.

Find the greatest velocity (in miles per hour) at which an engine of 192 horse-power can pull a train of 150 tons mass along the level, the frictional and other resistances to the motion being 16 pounds weight per ton. (A horse-power is 550 foot-pounds per second.)

12. State the laws which determine the velocities, after impact, of two smooth elastic spheres whose velocities before impact are known.

A smooth billiard ball strikes another (of equal size) which is at rest, and the direction of the centre of the former makes, just before impact, an angle of 30° with the line joining the centres of the two balls; find the tangent of the angle through which its direction of motion is deflected by the impact, the coefficient of restitution of the balls being 0·4, and the motion being supposed to take place upon a smooth billiard table.

VII. PART III. SECOND PAPER.

[*Full marks will be given for about* two-thirds *of this paper. The acceleration due to gravity may be taken to be* 32 *foot-second units.*]

1. Having given a focus, and one tangent of a parabola, and the length of the latus rectum, find the point of contact of the given tangent.

2. Prove that any tangent of a hyperbola cuts off from the asymptotes a triangle of constant area.

If the tangent at P meets an asymptote in L, and the perpendicular from the centre C on the tangent PL is equal to the perpendicular from L on CP, the asymptotes are at right angles.

3. The equation of a straight line being expressed in the form

$$\frac{x-a}{\cos\theta} = \frac{y-b}{\sin\theta}$$

interpret the geometrical meaning of either member of the equation.

Straight lines parallel to $y = kx$ are drawn to cut the two given lines $y = mx$ and $y = nx$; find the equation of the locus of the middle points of the parallel intercepts.

4. Find the equation of the two straight lines drawn through the point (1, 1) which are parallel respectively to the straight lines given by the equation $x^2 + 5xy + 2y^2 = 0$.

Find also the equations of the two diagonals of the parallelogram formed by the four lines.

5. Form the equation of a circle of radius a touching both the axes of co-ordinates, and lying in the quadrant in which both co-ordinates are positive.

Find the angles which a chord PQ passing through the origin O must make with the axes in order that OQ may be bisected at P.

6. Find the equation of the tangent to the parabola $y^2 = 4ax$ at the point $\left(\frac{a}{m^2}, \frac{2a}{m}\right)$.

Prove that, if two tangents are drawn from a point on the line $x = -4a$, their chord of contact subtends a right angle at the vertex.

7. Form the equation of the diameter of the ellipse $x^2/a^2 + y^2/b^2 = 1$ which is conjugate to the diameter $y = x \tan\theta$.

Chords AP, AQ parallel to the two diameters are drawn through the vertex (a, o). Prove that PQ makes with the axis of x an angle

$$\tan^{-1}\frac{2\tan\theta}{1-\frac{a^2}{b^2}\tan^2\theta}.$$

8. A rigid body is under the action of any number of forces in one plane. State the conditions of equilibrium.

A vertical post is held upright in contact with the ground by two cords, attached to pegs on the ground at equal distances c from the foot of the post, and attached to the post at heights a and b above the ground; the cords being on either side of the post in a vertical plane. Prove that the tangent of the angle which the reaction of the ground makes with the vertical cannot exceed $\frac{1}{2}c(a-b)/ab$.

9. Two particles of given masses are moving in a plane with given velocities in given directions. Determine the velocity of their centre of mass.

If the masses of the particles are m and m', and the velocity of m relative to m' is of magnitude v and makes an angle α with the line drawn from m' to m, determine the magnitude and direction of the velocity of m relative to the centre of mass.

10. A body which weighs 10 lbs. is attached to a cord passing over a fixed smooth pulley, and the cord is drawn over the pulley at a uniform rate of 3 ft. per sec. What is the magnitude of the tension of the cord?

If the pulley is rigidly attached to a moving platform, which is ascending with an acceleration of 4 foot-second units, and the cord is drawn over the pulley as before, what change is made in the tension?

11. Two particles of equal masses are attached to the ends of a light string of length l; one of them, A, is placed on a smooth table and the other, B, hangs just over the edge, the string being just tight. Find the magnitudes and directions of the velocities of the two particles (1) just before A leaves the table, (2) just after.

12. A particle is projected from a given point with a given velocity in a given vertical plane; find the envelope of the possible paths.

A shot is fired from a gun at the top of a cliff of height h with a velocity u ft. per sec. Prove that, if the range measured from the foot of the cliff is as great as possible, the elevation α is given by the equation

$$\cos 2\alpha = gh/(gh+u^2).$$

MATHEMATICAL EXAMINATION PAPERS

FOR ADMISSION INTO

Royal Military Academy, Woolwich,

NOVEMBER, 1900.

OBLIGATORY EXAMINATION.

I. PART I. FIRST PAPER.

[*Great importance is attached to accuracy.*]

1. Express as a fraction in its lowest terms

$$\{\tfrac{9}{11} \text{ of } 17\tfrac{3}{7} + \tfrac{18}{35} \text{ of } 116\tfrac{4}{45}\} \times \{\tfrac{11}{1700} \text{ of } 144\tfrac{1}{6}\}.$$

2. Calculate to five places of decimals (so that the error is less than 0·000005) the value of the expression

$$\frac{14\cdot52}{83\cdot1} + \frac{1\cdot575}{13\cdot85}.$$

3. A rectangular field is 330 yards in length and 188 yards in breadth; find the number of acres in the field, and the money which would be obtained by selling half the field for £17. 4*s.* 6*d.* an acre, and the other half for £21. 15*s.* 6*d.* an acre.

4. A person, having four thousand pounds to invest, places £2000 in $2\frac{3}{4}$ per cent. consols at $98\frac{1}{2}$, and £2000 in India three per cents. at $98\frac{5}{8}$; find, to the nearest penny, the income thereby obtained.

5. A railway train, 73 metres in length, is travelling at the rate of 60 kilometres an hour on the up line, and another train, 102 metres in length, is travelling in the opposite direction, on the down line, at the rate of 40 kilometres an hour; find the time occupied by the trains in passing each other.

If the trains were travelling in the same direction, what would be the time occupied by the faster train in passing the slower train?

6. A clockmaker was asked how many clocks he had, and replied: I have more than 500 but less than 1000; if I count them four by four, or five by five, or six by six, there remains one over, if I count them by sevens none remain over. How many clocks had he?

7. Show that

$$(ax+by+cz)^2+bc(y-z)^2+ca(z-x)^2+ab(x-y)^2$$
$$=(a+b+c)(ax^2+by^2+cz^2).$$

8. Determine c so that the equation in x

$$\frac{x-3}{x-4}+\frac{x-4}{x-3}=\frac{2x-c}{x-5}$$

has zero for a root. With this value of c solve the equation completely.

9. Solve the simultaneous equations

$$x(b-c)+by-c=0, \quad y(c-a)-ax+c=0.$$

Show that the values which satisfy the above equations will also satisfy the equation $ax-by+a-b=0$.

10. Prove the Index Law, namely that $a^m \times a^n = a^{m+n}$, when m and n are positive integers.

If the Index Law be assumed to hold good for all values of the indices, find what meanings must be assigned to a^0, a^{-m}, and $a^{\frac{m}{n}}$.

Multiply $a^{\frac{2}{3}}-a^{\frac{1}{3}}b^{\frac{1}{3}}-2b^{\frac{2}{3}}$ by $2a^{\frac{2}{3}}-a^{\frac{1}{3}}b^{\frac{1}{3}}-b^{\frac{2}{3}}$, and divide the product by $2a^{\frac{2}{3}}-3a^{\frac{1}{3}}b^{\frac{1}{3}}-2b^{\frac{2}{3}}$.

11. Find the arithmetic mean of x and y.

If a, b, c, are in arithmetical progression, show that $2a-3b$, $2b-3c$, and $3b-4c$ will also be in arithmetical progression.

Show that $3(1+3+5+\ldots+99)=101+103+105+\ldots+199$.

12. Write down expansions of $\frac{1}{\sqrt{1+x^2}}$, one of them valid when $x<1$, the other when $x>1$.

Use one of the expansions to find its value to the nearest millionth when $x=10$.

II. PART I. SECOND PAPER.

N.B.—*In questions on Geometry ordinary abbreviations may be employed, but the method of proof must be geometrical. Proofs other than Euclid's must not violate Euclid's sequence of propositions. In the absence of special directions to Candidates, any of the propositions within the limits prescribed for Examination may be used in the solution of deductions.*

1. (i.) Define a *parallelogram.*

(ii.) If, of the sides of a quadrilateral, one pair is parallel and the other pair equal, prove that any angle is either equal or supplementary to the opposite angle.

2. (i.) Prove that the exterior angle of a triangle equals the sum of the two interior and opposite angles.

(ii.) In a certain quadrilateral the sum of two internal adjacent angles is 200°. Find, in degrees, the size of the obtuse angle contained by the bisectors of the two remaining internal angles of the quadrilateral.

3. (i.) Show, with accompanying proof, how to construct a square equal to a given rectilinear figure.

(ii.) By means of a figure carefully drawn to scale, *apply the construction in* (i.) to find, as accurately as you can, the lengths of the sides of a rectangle whose area is 9 sq. in. and whose perimeter (the sum of the sides) is 14 inches. A proof is not required.

4. Prove that the sum of two opposite angles of a quadrilateral inscribed in a circle equals two right angles.

5. (i.) Give, without the proof, the construction for drawing a tangent to a circle from an external point.

(ii.) P, C and L denote respectively a given point, circle and unlimited straight line, P and C lying on the same side of L. Find, with proof, a point Q in L such that PQ and a tangent from Q to C (not in the same line with PQ) make equal angles with L. Only one solution is required.

6. Show how to circumscribe a circle about a given triangle ABC.

If O is the centre of this circle, and the lines AO, BO, CO produced meet the circumference in A', B', C', show that the triangle $A'B'C'$ is in every respect equal to ABC.

7. Show how to find a mean proportional between two given right lines.

O is a point outside a given circle; through *O* is drawn any right line meeting the circle in *A* and *B*; state (without proof) why the mean proportional between *OA* and *OB* is the same for all lines drawn through *O*, and exhibit this mean proportional.

8. Prove that the area of a trapezium is equal to half the sum of the parallel sides multiplied by the perpendicular distance between them.

The parallel sides of a trapezium are 9 and 30 feet long, and the other sides are 17 and 10 feet long; find its area.

9. On the base of a hemisphere of radius r is constructed a right cone whose volume is equal to that of the hemisphere; what is the height of the cone?

10. A spherical shell 1 foot in external diameter and 2 inches thick is made of metal having a mass of 504 pounds per cubic foot; find (to the nearest pound) the mass of metal in the shell. (The circumference of a circle is 3·14159 times the diameter.)

III. PART I. THIRD PAPER.

1. State the most approximate value you know for the numerical quantity π, and compare it with the value of the fraction $\frac{355}{113}$ expressed as a decimal.

What is meant by an angle π, and why?

Calculate the value of a radian expressed in seconds.

2. Express all the other trigonometric functions of A in terms of $\sin A$. Why must the double sign $\pm$ be prefixed to all the square roots occurring in your expressions?

Write down a general formula for all angles whose sine is equal to $\frac{1}{2}$ expressed *either* in degrees *or* in radians.

3. Simplify

$$(\sec\theta - \cos\theta)(\operatorname{cosec}\theta - \sin\theta)(\tan\theta + \cot\theta)$$

and find a positive angle θ less than 180° satisfying the equation

$$1 - \cos\theta = \sqrt{3}\sin\theta.$$

4. Write down and prove the formula giving $\tan(A+B)$ in terms of $\tan A$ and $\tan B$.

The perpendicular from the vertex of a triangle on the base is 6 inches long, and it divides the base into segments which are 2 and 3 inches long respectively. Find the tangent of the vertical angle of the triangle.

5. An observer wishing to determine the length of an object in the horizontal plane through his eye, finds that the object subtends the angle α at his eye when he is in a certain position A. He then finds two other positions B and C where the object subtends the same angle α. Express the length of the object in terms of the sides of the triangle ABC and of the angle α.

---

6. What is meant by the supplement of an angle? What is the supplement of an angle of 330°? Draw a figure showing the angle and its supplement, and prove that the sines of these two angles are algebraically equal.

Find $\operatorname{cosec} 630^\circ$.

7. Prove the formula

$$a^2 = (b + c - 2\cos\frac{A}{2}\sqrt{bc})(b + c + 2\cos\frac{A}{2}\sqrt{bc}).$$

Apply it to find the side a of a triangle when $b = 132{\cdot}5$ feet, $c = 97{\cdot}32$ feet, $A = 37^\circ\ 46'$, as accurately as the tables permit.

8. Find an expression for the difference in area of the two triangles which can be drawn having two given sides a, c and a given angle A opposite the smaller side a.

9. Explain the meaning of $\log_a m$.

Prove that $\log_a m \times \log_m a = 1$.

Without using the tables find the characteristics of (1) $\log_7 15914$, (2) $\log_8 0{\cdot}00187$.

10. Throw the expression

$$\frac{5\sin\theta + 3{\cdot}584\cos\theta}{5\sin\theta - 3{\cdot}584\cos\theta}$$

into a form suitable to logarithmic calculation when different values of θ are introduced, and use your form to evaluate the expression when $\theta = 71^\circ\ 59'$ correct to three places of decimals.

IV. PART II. First Paper.

[*Full marks will be given for about* three-quarters *of this paper.*]

1. (i.) Simplify

$$\frac{x^2y^2(x^2-y^2)+y^2z^2(y^2-z^2)+z^2x^2(z^2-x^2)}{xy(x-y)+yz(y-z)-zx(z-x)}.$$

(ii.) If the sum of a, b, and c be p, show that the sum of $a(b+c)$, $b(c+a)$, and $c(a+b)$ cannot be greater than p^2.

2. Trace the graph of the function x^3-4x+1. (The *graph* is the locus of a point whose distance from one straight line is x, and whose distance from a perpendicular straight line is the corresponding value of x^3-4x+1.)

From your diagram show that the given function vanishes for three values of x, and find between what pair of integers each of these values lies.

3. ABC is a triangle; BA, CA are produced through A to points D, E, so that ED and BC are parallel. Prove that the circles ABC, ADE touch one another.

4. Prove that the volume of a tetrahedron is one-third that of the prism on the same base and of the same altitude.

In a regular tetrahedron prove that the perpendicular from a summit on the opposite face is three times the perpendicular from the centre of gravity of a face upon any other face.

5. Express the tangent of an angle in terms of the cosine, explaining the ambiguity of sign.

Having given $\tan\theta=\sqrt{15}+2\sqrt{3}$, $\cos\alpha=\frac{1}{4}$, and that α is an acute angle, show that $\tan(\pi-\alpha-\theta)=\sqrt{15}-2\sqrt{3}$.

6. Investigate some expression for the radius of the circle circumscribing a triangle, and write down any other expressions you know.

A man is on the perimeter of a circular field, and wishing to know its diameter, he selects two points in the boundary a furlong apart, which at a third point, also in the boundary, subtend an angle of 164° 43′. Find the diameter to the nearest foot.

7. Two intersecting straight lines are respectively parallel to two other intersecting straight lines. Show that the plane containing the first pair is parallel to that determined by the second pair, or else coincides with it.

8. Define a trihedral angle, and show that the sum of any two of the plane angles at the summit is greater than the third.

Find the locus of a point which is equidistant from three lines which meet at a point.

9. T is a point exterior to a conic section, and L, N are its orthogonal projections upon a fixed focal chord and the directrix respectively; if the ratio $SL : TN$ be equal to the eccentricity, show that T lies on one or other of the tangents at the extremities of the focal chord.

Deduce from this theorem a construction for drawing tangents to a conic from a given point T.

10. Take two points A and B an inch apart. Three conics have A for vertex, B for focus, and eccentricities 0·2, 1, 5 respectively. Determine a few points on each curve and draw the curves freehand.

11. Show that the locus of the middle points of any system of parallel chords of a conic is a straight line which meets the directrix on the straight line through the focus at right angles to the chords.

If a tangent be drawn parallel to any chord of a conic, prove that the portion of it terminated by the tangents at the ends of the chord is bisected at its point of contact.

12. Prove that one quarter of the rectangle between any focal chord and the latus rectum is equal to the rectangle between the segments of the chord determined by the focus.

Enunciate some other proposition concerning the length of the latus rectum which is true for the general conic.

V. PART II. SECOND PAPER.

[*Full marks will be given for about* three-quarters *of this paper. The diagrams supplied are to be themselves used, and not pricked off into your book.*

A body in a vacuum falls to the earth with an acceleration of 32 *foot-second units.*]

1. Show how to resolve a force, represented by a given segment of a given straight line, into components, whose lines of action are parallel to two given straight lines.

If the direction of one of the components of the given force is given, but not its magnitude, and the magnitude of the other component is given, but not its direction, construct the magnitude of the former component and the line of action of the latter.

2. State (without proof) the conditions of equilibrium of a system of forces acting in one plane at a point.

Two small smooth rings of equal weight are free to slide on a thin circular wire, which is fixed in a vertical plane, and are connected by an inextensible thread, the length of the thread being less than the diameter of the circle. Prove that, if the system rests with the thread in a state of tension, the line of the thread must be horizontal.

3. *AB* represents a uniform bar of weight 11 lbs. and length 3·2 inches which rests at an angle of 50° on a rough horizontal plane at *A* and against a smooth peg at *C*, distant 2¼ inches from *A*. Find the magnitude of the reaction of the plane on the bar.

4. Prove that the sum of the moments of two like parallel forces about any point in the plane through their lines of action is equal to the moment of their resultant about the same point.

A uniform bar, of weight 5 lbs. and length 22 inches, is supported by two vertical cords attached to its ends, and two weights of 5½ lbs. each are attached to it at points distant 8 inches and 18 inches from the same end. Find the tensions of the two cords.

5. A circular disc, lying on a smooth table, has strings attached at *A* and *B*, which are pulled with forces represented by the lines *AC* and *BD* (an inch representing one poundweight). A third string is to be attached and pulled so as to keep the slab at rest. Determine a point *E* at which it may be attached, and represent by a line *EF* the magnitude and direction of the pull.

Also write down the magnitude of the pull.

$AC = 2{\cdot}44$ inches, $BD = 3{\cdot}18$ inches, and the angle between *AC* and *BC* $= 65°$.

6. Draw a figure of that system of pulleys in which the same cord passes round all the pulleys, and find the mechanical advantage of the system.

If a force of 50 poundsweight supports an attached mass of 220 lbs., and a force of 25 poundsweight supports an attached mass of 95 lbs., what force will be required to support an attached mass of 315 lbs., and what is the weight of the lower block?

7. A body moves from rest with uniform acceleration. From *O* is measured along *OA* a length *OP* representing any time elapsed. *OB* is drawn at an angle of $22\frac{1}{2}°$ with *OA* so that $\tan BOA$ represents the acceleration, and *PQ* is drawn perpendicular to *OA* to cut *OB* in *Q*. Show that *PQ* represents the speed at the time *OP*, and the area *OPQ* the space described in that time.

If the scales are chosen so that half an inch along *OA* represents a second, and an inch perpendicular to *OA* represents a speed of one foot per second, find the speed acquired in ten seconds.

8. Prove that, when a point moves with uniform acceleration, its average velocity over any interval of time is half the sum of its velocities at the beginning and end of the interval.

A train runs from rest at one station and stops at the next. During the first quarter of the journey the motion is uniformly accelerated, and during the last quarter it is uniformly retarded, the middle portion being performed at a uniform full speed. Prove that the average speed of the train is $\frac{2}{3}$ of the full speed.

9. Two bodies of equal mass, each *m* lbs., are attached to an inextensible thread; one of them is placed on a smooth table, and the other hangs over the edge; find the acceleration.

When the system has been moving for one second, the body on the table not having reached the edge, another body of equal mass, which was at rest on the table, becomes attached to it in such a way that the compound body moves in the line of the thread. Find the impulse of the tension in the thread.

10. A body which weighs 10 lbs. is observed to move horizontally with an acceleration of 4 foot-second units; find the magnitude and direction of the force applied to it.

If the body hangs from the roof of a railway carriage by equal strings attached to two points one directly in front of the other so that the strings are at right angles, and if the acceleration of the train is 4 foot-second units in the direction of motion, find the ratio of the tensions of the strings.

11. A particle is projected from a given point with velocity V, in a direction making an angle α with the horizontal; find the range on a horizontal plane through the point of projection.

A fielder 300 feet from the wicket throws in the ball with a speed of 100 feet per second and at an elevation of 30°. Will the ball return to the level of his hand before or after passing the wicket? The air-friction may be neglected.

12. A weight is attached to a cord of given length, of which the upper end is fixed, and the cord is held so as to make with the vertical an angle α. Find the velocity with which the weight must be projected horizontally in order that it may describe a horizontal circle uniformly.

VOLUNTARY MATHEMATICS.

VI. PART III. FIRST PAPER.

[*Full marks will be given for about* two-thirds *of this paper.*]

1. C is the centre of an ellipse, and A, B extremities of the major and minor axes. If PN be the ordinate of P, and PG the normal, prove that

$$NG : CN :: CB^2 : CA^2.$$

2. Define an hyperbola in reference to a focus and a directrix, and thence establish the property enjoyed by the focal vectors of any point on the curve.

Given two points on a parabola and the direction of the axis, find the locus of the focus.

3. Investigate the equation to a straight line in the form

$$x \cos \alpha + y \sin \alpha = p.$$

Show that the three straight lines

$$x \cos \alpha + y \sin \alpha = p$$
$$x \cos \beta + y \sin \beta = q$$
$$x \cos \gamma + y \sin \gamma = r$$

will pass through the same point if

$$p \sin (\beta - \gamma) + q \sin (\gamma - \alpha) + r \sin (\alpha - \beta) = 0.$$

This condition being satisfied, find the co-ordinates of the point.

4. Prove that the equation

$$p\{(x-\alpha)^2 + (y-\beta)^2 - r^2\} + q\{(x-\gamma)^2 + (y-\delta)^2 - s^2\} = 0$$

denotes a circle.

Find its radius, and show that its centre divides the line joining the centres of the circles

$$(x-\alpha)^2+(y-\beta)^2=r^2, \ (x-\gamma)^2+(y-\delta)^2=s^2$$

in the ratio $p:q$.

5. Under what circumstances does an equation of the second degree denote a parabola?

Find the direction of the axis of the parabola

$$\cos^2\alpha\, x^2+\sin 2\alpha\, xy+\sin^2\alpha\, y^2+2gx+2fy+c=0$$

and the co-ordinates of its vertex.

6. Two concentric circles are described about a fixed point C as centre, and any radius vector CPQ cuts them in P and Q. If a rectangle $PAQB$ be described about PQ as diagonal, so that PA, PB are in fixed directions, show that as the direction of CPQ varies A and B move upon ellipses.

Show the ellipses in a figure.

7. Tangents are drawn from the point (p, q) to the conic $ax^2+by^2=1$.

(i.) Find the equation of the tangents.

(ii.) Determine the conditions that must be satisfied by the numbers p, q, if the tangents are to include an angle α, and give a geometrical interpretation.

8. Give the complete specification of a couple, and show how couples may be compounded.

A circular disc of weight W and radius a is suspended horizontally by 3 equal vertical strings of length b attached symmetrically to the perimeter. Find the magnitude of the couple required to keep it twisted through a given angle θ.

9. A weight can be sustained upon a rough inclined plane by a force P acting along the plane or by a force Q acting horizontally; show that the weight is $\dfrac{PQ}{(Q^2\sec^2\phi-P^2)^{\frac{1}{2}}}$ where ϕ is the angle of friction.

10. Define relative velocity and relative acceleration. A stone is projected from the top of a tower with velocity v and at an angle of elevation α, and, at the same instant, another stone is dropped from the same place. Find the relative velocity and acceleration of the first stone with respect to the second.

11. Define the poundal, the dyne, and the horse-power. Find the speed of a train, in miles per hour, given that the locomotive of 50 horse-power works against a resistance of 10,000 poundals.

12. Calculate the length of a seconds pendulum at a place where a body falls to the earth with an acceleration of 32·19 foot-second units.

Given that the force of gravity varies inversely as the square of the distance from the earth's centre, find an approximate formula for the number of seconds a day lost by a seconds pendulum when taken to a height h, which is small compared with the earth's radius.

If h be 5 miles, and the earth's radius 3956 miles, show that the loss is about 109 seconds a day.

VII. PART III. SECOND PAPER.

[*Full marks will be given for about* two-thirds *of this paper. The acceleration due to gravity may be taken to be* 32 *foot-second units.*]

1. From a point T two tangents TP and TQ are drawn to touch a parabola, whose focus is S, at P and Q, and T' is any point on TS produced beyond S. Prove that the angles PTQ, PST', QST' are all equal.

2. From the foci S and S' of an ellipse perpendiculars SY and $S'Y'$ are drawn to the tangent at any point P. Prove that Y and Y' lie on the auxiliary circle.

If the perpendicular from the centre on the tangent at P and a focal radius SP (produced if necessary) meet in R, then SR is equal to half the major axis.

3. Find an expression for the length of the perpendicular from the point (h, k) to the straight line $ax+by+c=0$.

Find the locus of a point which moves so that the perpendiculars drawn from it to the two straight lines

$$3x+4y=5, \quad 12x-5y=13$$

are equal.

4. Explain how a single equation can represent two straight lines.

Form a single equation to represent the two straight lines

$$3y-2x=1, \quad 5y+4x=2.$$

Also find the equation of the two straight lines joining the origin to the points where these lines are met by the line $y=\frac{2}{11}$.

5. Two circles are drawn to touch both the lines $y=0$, $y=x\tan 2a$, and to pass through the point (h, k); prove that the distances of the points of contact from the origin are the roots of the equation

$$\xi^2-2(h+k\tan a)\xi+h^2+k^2=0.$$

Find the equation of the common chord of the two circles.

6. Find the equation of the tangent to the parabola $y^2=4ax$ at the point (x', y').

From a fixed point (h, k) a line is drawn at right angles to the tangent to the parabola at P, and meeting the diameter through P in Q. Prove that the locus of Q for different positions of P is the rectangular hyperbola

$$y(x+2a-h)=2ak.$$

7. Form the equation of the chord of an ellipse $\frac{x^2}{a^2}+\frac{y^2}{b^2}=1$ joining the points of which the eccentric angles are θ and ϕ.

Prove that, if P is the point for which the eccentric angle is $\frac{1}{2}(\theta+\phi)$, and the line joining P to either focus meets the chord in Q, then

$$QP=a\left(1-\cos\frac{\theta-\phi}{2}\right).$$

8. Prove that, if three forces in one plane acting on a rigid body are in equilibrium, their lines of action meet in a point or are parallel.

A uniform square lamina is supported, so that its plane is vertical and one of its corners is in contact with a fixed smooth plane of inclination α, by a force applied at the opposite corner in a direction parallel to the line of slope of the inclined plane. Prove that the diagonal which passes through the point of contact makes with the vertical an angle 2α.

9. Two particles of masses, m, m', are connected by an inextensible thread; m is placed on a rough inclined plane, and a thread tied to it lies along a line of slope, passes over a pulley at the top of the plane, and supports m' which hangs free. Find the condition that m should move (1) upwards, (2) downwards, and the acceleration in each case. [Take μ for the coefficient of friction.]

10. A body which weighs one ton is supported by a wire rope which weighs two pounds per foot. Find the tension at the upper end of the rope when the body is just above the bottom of a mine 200 feet deep.

Find also the work done in lifting it gradually to the surface.

11. A smooth sphere lying at rest on a horizontal table is struck by an equal sphere moving on the table so that just before the impact the direction of its motion makes an angle α with the line of centres. Taking e for the coefficient of restitution, find the directions of motion of both spheres after impact.

12. Prove that the velocity of a projectile at any point is that due to falling from the directrix of its path.

Prove also that, if V is the velocity of projection and α the initial angle of elevation, the equation of the path referred to horizontal and vertical axes through the point of projection is

$$gx^2=2V^2\cos^2\alpha\,(x\tan\alpha-y).$$

MATHEMATICAL EXAMINATION PAPERS

FOR ADMISSION INTO

Royal Military Academy, Woolwich,

JUNE, 1901.

OBLIGATORY EXAMINATION.

I. ARITHMETIC. (1.)

(To vulgar fractions and decimals.)

[N.B.—*You are requested to put the* number *to each question, and to send up the* working *as well as the answers.*]

1. Multiply £728. 15*s*. $4\frac{3}{4}$*d*. by 8.
2. Divide 3 tons 17 cwt. 3 qrs. by 7.
3. How many seconds are there in 1 day 17 hours 11 minutes?
4. How many kilometres and metres are there in 5184000 centimetres?
5. What are the prime factors common to 374 and 204?
6. Find the L.C.M. of 8, 15, 25, 36.
7. What fraction with denominator 416 is equal to $\frac{11}{8}$?
8. Add together $4\frac{11}{24}$ and $9\frac{7}{36}$.
9. Subtract $8\frac{3}{14}$ from $11\frac{2}{21}$.
10. Multiply $7\frac{3}{5}$ by $1\frac{1}{38}$.
11. Divide $8\frac{3}{14}$ by $8\frac{4}{21}$.

W.P. 2K

12. Add together 55·78, 4·016, 0·054, 19·7.

13. Subtract 8·715 from 11·2.

14. Multiply 7·04 by 0·175.

15. Divide 0·3321 by 7·38.

16. Express [illegible] as a decimal.

17. Express as a fraction in its lowest terms the ratio of 4 litres 55 centilitres to 7 hectolitres.

18. Express in miles, furlongs, poles, yards and the decimal of a yard, 0·457 of 3 miles.

19. Find the cost of 84 articles at 14*s.* 5¾*d.* a score.

20. Find the value of 6 poles 3 yds. 1 ft. 7 ins. at 16*s.* 6*d.* a pole.

21. If 49 bushels of wheat cost £16. 1*s.* 5*d.*, what would 84 bushels cost at the same rate?

22. If 25 people spend £1624 in 9 months, how much would 45 people spend in half as long again at the same rate of spending?

23. How many quarters are there in 13 per cent. of 3 tons 15 cwt.?

24. What percentage is 510 grams of 17 kilograms?

25. How many cubic metres are there in a block 15 metres long, 5 metres wide and 3 metres high?

II. ARITHMETIC. (2.)

[*In order to obtain full marks arithmetical methods of solution are required. All working must be clearly shown.*]

1. Employ an abbreviated method to divide 33·82289 by 66·4684 to six places of decimals (so that the error is less than 0·0000005).

2. A bicycle was sold at a loss of 40 per cent. on the cost price, and a second bicycle was bought with the proceeds when 5 guineas had been added. This second bicycle was sold at a loss of 35 per cent., and a third bicycle was bought with the proceeds after the addition of £5. 15*s*. 3*d*. This third bicycle cost £16; what was the price of the first?

3. A hollow ball, of external diameter 6 centimetres and of thickness half a centimetre, weighs 162 grams when it is filled with liquid. Find to the nearest hundredth of a gram the weight of a cubic centimetre of the liquid, a cubic centimetre of the substance of the ball weighing 2 grams. (If a sphere is inscribed in a cube, then volume of sphere : volume of cube : : 11 : 21.)

4. Tithe Rent Charge in 1901 is of such value that every nominal £100 is really worth only £66. 10*s*. 9¼*d*. Find the amount of Tithe Rent Charge which is worth £1. 14*s*. 4*d*. less than its nominal value.

5. An investor placed a sum of money in a stock *A*, and a sum £1000 greater in a stock *B*, getting £30 a year more from the second than from the first. Had he invested the whole in *A*, he would have obtained £5 a year more than if he had invested it in *B*. Find what income would arise from an investment of £2000, half in *A* and half in *B*.

6. Write down the nine numbers which immediately precede 81 and the nine which immediately follow, and show that their sum is exactly divisible by 9. Would the sum still be divisible by 9, if instead of a group of 9 numbers on either side of 81, a group of five, six, or any other amount were chosen? Give your reasons.

III. PART I. First Paper.

[N.B.—*In questions on Geometry ordinary abbreviations may be employed, but the method of proof must be geometrical. Correct demonstrations, whether those of Euclid or not, will be accepted. In the absence of special directions to Candidates, any of the propositions within the limits prescribed for Examination may be used in the solution of deductions.*]

1. Show that if two triangles ABC and XYZ have
side AB = side XY, side BC = side YZ, angle B = angle Y
the triangles are equal in every respect.

Two quadrilaterals $ABCD$ and $XYZW$ have

$$AB = XY,\ BC = YZ,\ CD = ZW,\ B = Y,\ C = Z.$$

Show that they are equal in every respect.

2. Show that any two sides of a triangle are together greater than the third.

Give in substance the enunciation of propositions from which you know that the sum of the squares on two sides of a triangle may be less than, equal to, or greater than, the square on the third side.

3. Show that the opposite sides of a parallelogram are equal.

ABC is a triangle, E the middle point of AB, D the point of trisection of CA nearer to C; BD and CE cut in O. Show that DO is a quarter of DB.

4. Show that the angle at the centre of a circle (whether less than, equal to, or greater than two right angles) is twice the angle at the circumference which stands on the same arc.

$ABCD$ is a quadrilateral of which the angles at B and D are right. Show that the difference between the angles BAC and DAC is equal to the difference between BCA and DCA. What is the difference if the diagonals AC and BD cut at an angle of $60°$?

5. A, B, C, D are four points lying in this order on a straight line. $AC = a$ inches, $CD = b$ inches, $BC = 1$ inch. A perpendicular at C to the line cuts in E the circle on BD as diameter. EFG passes through the

centre of the circle drawn on AC as diameter, cutting the circumference in F and G. Show that the area of the rectangle $EF.EG$ is b square inches.

Show also that the number of inches in the length of EF is a root of the equation $$x^2+ax-b=0.$$

6. With your instruments draw a circle 5 centimetres (or, if you prefer it, 2 inches) in radius, and inscribe in it a regular hexagon.

State your construction and prove that all the angles of the hexagon are equal.

7. Show that the area of a triangle is $\frac{1}{2}ab$ square inches if the base is a inches long and the altitude b inches, and a and b are whole numbers.

Assuming the expression true for all values of a and b show hence that triangles of the same altitude have areas proportional to their bases.

8. Prove that a line parallel to the base of a triangle divides the sides in the same ratio.

Show (with proof) how to divide a given straight line into 3 equal parts.

9. Take two points, A and B, 9 centimetres (or, if you prefer, $3\frac{1}{2}$ inches) apart. With your instruments find any 5 points which are twice as far from A as from B.

State (without proof) the nature of the locus on which the five points must lie.

10. The ground beside a level road rises 1 foot in 3 feet (measured along the slope). How much earth must be removed to leave a level square surface measuring 70 feet along the road and 70 feet back from the road, the sides of the excavation being vertical? Give the result in cubic yards to the nearest integer.

11. Show that the area of a road bounded by two concentric circles is the breadth of the road multiplied by the arithmetic mean between the lengths of the boundaries.

A circular road is 20 metres wide and a kilometre long (measured along its central line). Find its area in square metres.

12. Find in degrees the angle of a regular pentagon, and from the tables find the ratio of the radius of the inscribed circle to the side.

The following construction is sometimes used to find the centre of the circle inscribed in the regular pentagon on AB as side:—Draw a quadrant of a circle with AB as one bounding radius. Let the perpendicular bisector of AB cut the chord and the arc of the quadrant in C and D. The middle point of CD is taken as centre. By how much per cent. is the radius in error?

WOOLWICH ENTRANCE EXAMINATION.

IV. PART I. SECOND PAPER.

1. Find the values of

$$1+x+\frac{x^2}{2!}+\dots\frac{x^n}{n!}+\dots$$

to the nearest hundredth when $x=1$ and when $x=2$.

Show that when $x=2$ every term after the 8th is less than $0{\cdot}01$.

2. Solve the equations

$$y+\sqrt{2}\,.\,x=6, \quad x=\sqrt{2}\,.\,y$$

giving y accurately, and x to the nearest tenth.

3. Show that if x^4+1-kx^2 vanishes when $x=a$, it vanishes also when x is equal to $\frac{1}{a}$ or $-a$ or $-\frac{1}{a}$.

Find the values of x for which it vanishes when $k=4{\cdot}25$.

4. For what value of x is

$$x^2-3x+2=(x-3)(x-4)?$$

For what values of a and b is

$$x^2-3x+2=(x-a)(x-b)$$

for all values of x?

5. If t is proportional to $pv^{\frac{3}{2}}$, and $t=224$ when $p=28$ and $v=16$, what is the value of t when $p=32$ and $v=25$?

6. Factorise the expression

$$2x^2+3xy+y^2-6x-4y+4$$

and solve completely the equations

$$2x^2+3xy+y^2-6x-4y+4=0$$
$$2x^2+4xy+y^2-4x-5y+2=0.$$

7. Having given $10^x=2$ calculate 10^{2x}, 10^{4x}, 10^{8x}, and 10^{16x}, and hence show that x must lie between $\frac{4}{16}$ and $\frac{5}{16}$.

Show (without using the tables) that $\log_{10}2$ lies between $0{\cdot}25$ and $0{\cdot}32$.

8. State the binomial theorem as generally as you can. Prove its truth for positive integral indices.

Give the first five terms of the expansion of $(1+x)^{19}$, showing each coefficient multiplied out.

9. Construct with ruler and compass an acute angle whose tangent is 3·5.

By measuring your figure, or otherwise, find the sine and the cosine of the angle to the nearest tenth. (You are at liberty to make further constructions if you wish.)

10. Find an expression for $\cos\frac{\theta}{2}$ in terms of $\cos\theta$. In what cases is the negative value of the square root to be taken?

By means of the tables verify the relation when $\theta = 70°$.

11. My window in London faces due south. Opposite, 65 feet away, is a long flat-topped building over which just half the sun appears at noon at midwinter. Find the height of the building above my position. (At midwinter at noon the sun's altitude is 15°.)

12. Find an expression for the sine of half an angle of a triangle in terms of the sides.

The sides of a triangle are 3·712, 4·27, and 3·843 inches long. Determine the least angle and state which angle is the greatest.

V. PART II. FIRST PAPER.

1. Divide x^4+ax+b by x^2+px+q so as to show the quotient and the remainder.

What conditions must be satisfied if the division is exact? Show that one condition may be written

$$4bp^2=p^6-a^2.$$

2. Show that the sum of r terms of a series

$$a_1+a_2+a_3+a_4 \ldots$$

of positive quantities tends to a finite limit as r increases if $\frac{a_{m+1}}{a_m}<k$ for all values of m greater than a certain finite value, where k is a proper fraction.

Show that $1+nx+\ldots\frac{n(n-1)\ldots(n-r+1)}{1.2\ldots r}x^r\ldots$, where n and x are real, is convergent if $-1<x<1$.

3. Find the length of the hypotenuse of a right-angled triangle, which has the other two sides m^2-n^2 and $2mn$ centimetres long.

Show that if $m=n+1$ the hypotenuse differs from another side by 1 centimetre, and find the three sides when $m=13$, $n=12$.

4. Write down expressions for the whole surface, the curved surface, and the volume of a right cone.

Wine is poured into a conical wineglass till half the inner surface of the glass is below the surface of the wine. Show that the glass is a little more than 35 per cent. full.

5. Draw on the same diagram the graphs of the functions $\sin x$, $\cos x$ and $\sin x+\cos x$.

Derive from your figure the general solution of the equation

$$\sin x+\cos x=0.$$

6. Find the product of

4·177, 0·04177, 0·0004177, 4177000

and find the square root of $(0{\cdot}07346)^3$.

7. Define a regular tetrahedron and determine its altitude in terms of the length of an edge.

Given three points in a plane find a point without the plane equidistant from them.

8. A focus and corresponding directrix of a conic are given, as well as the distances of a point of the curve from the focus and the directrix. Show how other points on the curve may be found.

9. The distance between the foci of a conic is two inches and the eccentricity is one-half; construct the curve, giving a written explanation (without proof).

10. Prove that the straight lines joining the extremities of two focal chords intersect in the directrix.

VI. PART II. Second Paper.

[*Fourteen questions, and not more, are to be answered.*]

Gravitational acceleration = 32 foot-second units.

1. Three forces acting at a point can be represented in magnitude and direction by the sides of a triangle taken in order. Prove, from first principles, that the algebraic sum of the moments of the three forces about any point in their plane is equal to zero.

2. Four forces acting in a plane at a point O, are represented in magnitude and direction by the lines OA, OB, OC, and OD. $OA = 2$ ins., $OB = 3{\cdot}5$ ins., $OC = 1{\cdot}2$ ins., $OD = 4$ ins. $\angle AOB = 15^\circ$, $\angle AOC = 90^\circ$, $\angle AOD = 221^\circ$. An inch represents 10 pounds weight. Draw a diagram to scale to represent the forces and measure off from the diagram the value of the resultant OF and the $\angle AOF$.

3. Define the term centre of mass. Explain how the stability of a body with a plane base resting on a plane horizontal surface depends upon the position of its centre of mass.

4. A rectangular column, 20 feet high and 4 feet square in section, is not fixed to the ground but merely rests on it. What force acting perpendicular to a face at its centre would overturn it, the material weighing 150 lbs. per cubic foot?

5. Explain the "principle of work," as applied to machines. Apply it to find the relation between the power and weight on an inclined plane, where the power acts parallel to the plane.

6. A diagram represents a uniform gate, $6\frac{1}{2}$ feet long, weighing 200 lbs., fastened by hinges 8 inches long to a gatepost AB. The gatepost is fixed to the ground at A and supported by a rod BC, $5\frac{1}{4}$ feet long, to a point C in the ground distant 3 ft. 7 in. from A. What must be the tension of the rod when the gate is shut if there is no tendency to bend the post where it enters the ground?

7. Find the centres of mass of the triangles ABC and ADC, and of the quadrilateral $ABCD$. Mark the three points with the letters P, Q, R.

In the diagram $AB = 2{\cdot}2$ in., $BC = 4{\cdot}2$ in., $AC = CD = 4{\cdot}6$ in., and $DA = 1{\cdot}4$ in.

8. In a diagram OX, OY, two lines at right angles to one another, are divided into ¼ inch divisions; the divisions along OX represent seconds, those along OY represent feet per second of velocity. AB is a line inclined at 28° to OX, and cutting OY in A at 2½ divisions from O.

In the diagram the ordinate of any point on the line AB represents the velocity of a body moving in a straight line at a time represented by the abscissa. Find the distances traversed in 2, 4, 8, 16 seconds; and, using these numbers, draw on the diagram a curve showing the distance traversed by the body in various times.

[Take one division of the vertical scale to represent ten feet.]

9. Prove by means of a diagram, or otherwise, the formula $s = ut + \frac{1}{2}at^2$; s is the space described in time t by a body moving in a straight line with initial velocity u and acceleration a.

10. A stone, of mass ¼ lb., is thrown along the surface of a frozen lake with an initial speed of 30 feet per second, and travels 200 yards. Find the retarding force (supposed uniform) in poundals.

11. A toboggan slides down a smooth incline 100 feet long, which slopes 1 in 3 (measured along the slope). Find (1) the speed, (2) the momentum, when it reaches the bottom; the mass of the toboggan and rider being 160 pounds. Give your results to the nearest hundredth.

12. A sandbag, of mass 20 lbs., is let fall from a balloon which is rising with a (uniform) velocity of 10 ft. per second, when the balloon is 500 feet high. Find the impulse with which the bag strikes the ground.

13. How would you show experimentally that the weights of two bodies, *at the same place*, are proportional to their masses. Explain the importance of the words in italics.

14. Define the terms *work*, *power*.

A *horse-power* is 550 foot-pounds per second, a *watt* is 10^7 ergs per second. Show that a kilowatt is approximately equal to 1⅓ horse-power.

[1 pound = 460 grams; 1 foot = 30 centimetres; gravitational acceleration = 32 foot-seconds units.]

15. Every minute a pump raises 50 gallons of water 27 feet. A gallon of water weighs 10 lbs. Find in foot-pounds the work done in an hour, if half the work is wasted in friction.

16. A bullet is to be fired at a point 500 yards away in the same horizontal plane as the muzzle of the rifle. The muzzle velocity of the bullet is 2,000 feet per second; at what elevation should it be fired?

VOLUNTARY MATHEMATICS.

VII. PART III. FIRST PAPER.

[*Full marks will be given for about* two-thirds *of this paper.*]

1. Prove that the tangent at any point of a central conic section bisects one of the angles between the focal distances of the point.

Hence show that two conics which have the same two foci cut each other at right angles.

2. Find the centre and the principle axes of a given ellipse. Also state your construction.

[Three-fourths of the circumference of the ellipse was shown in a diagram.]

3. Prove that the area of a rectilineal figure, given by the co-ordinates of its vertices, depends upon the order in which the vertices are taken.

Obtain the area of the four-sided figure whose vertices are (taken in order) (0, 4), (2, 7), (7, 6), (9, 9); and the area of each of the triangles which make up the figure.

4. If (x_1, y_1) and (x_2, y_2) be the co-ordinates of two points P_1, and P_2 respectively, show that ax_1+by_1+c and ax_2+by_2+c will be of the same sign if P_1 and P_2 are on the same side of the line whose equation is $ax+by+c=0$; but that the two expressions will be of opposite signs when P_1 and P_2 are on opposite sides of the line.

5. The equation $x^2+y^2+2\lambda x+c=0$ represents a series of circles, one circle for each value of λ. Prove that, if c is negative, each of these circles passes through two fixed points, the same for all the circles. If c is positive find the centres of the particular circles of the systems whose radius is 0.

Determine the equation of a circle, which cuts every circle of the system at right angles.

6. Reduce the equation $35x^2-74xy+35y^2+x+13y=0$ to the forms $ax^2+2hxy+by^2=c$, and $(ax+by+c)(a'x+b'y+c')=C$.

Find the co-ordinates of the centre of the conic determined by the equation.

7. Define pole and polar with respect to a conic. Find the equation to the polar of a point (h, k) and the co-ordinates of the pole of a line $(y = mx + c)$ with respect to the conic $b^2x^2 + a^2y^2 - a^2b^2 = 0$.

Find the angle between two lines which intersect in a focus of the conic $b^2x^2 + a^2y^2 - a^2b^2 = 0$, each line passing through the pole of the other.

8. A circular plate of radius a feet is fixed in a vertical plane. Upon it (and in its plane) are placed two uniform heavy rods AB, BC, jointed at B, each being $2b$ feet in length and M pounds in mass. Find an equation for θ, the angle which either rod makes with the vertical in the position of equilibrium, and express in terms of M, θ the pressure upon the plate and the reaction at the joint. Friction is to be neglected.

9. A heavy particle of mass one pound rests upon a rough plane, whose angle of inclination is 19°. The angle of friction is 20°. What force acting at an angle of 48° to the inclined plane is just sufficient to move the particle up the plane? Determine also the direction for which the requisite force is least.

10. Define acceleration, force, work, and write down the dimensions of each in terms of the units of length (L), time (T), and mass (M). If the unit of force is 100 poundals, the unit of length 1 inch, and the unit of velocity 1 foot per minute, find the units of mass and time.

11. What are kinetic energy, potential energy, and work? How are they measured, and what are the fundamental relations between them?

A ball weighing 5 ounces moving at 1000 feet a second pierces a plate and moves on with a velocity of 400 feet a second. What kinetic energy would have been lost if the plate had been only half its actual thickness? (Assume the work done in piercing a plate to be proportional to the thickness.)

12. A smooth tube is bent into the form of Γ, so as to have a sharp corner. Two particles B and C joined to a third particle A by separate strings of lengths b feet, $b+c$ feet respectively are placed in the horizontal tube. Initially A is placed at the top of the vertical tube, while B and C are as far from A as the strings permit. The three particles being of equal mass, find how many seconds will have elapsed before B reaches the top of the vertical tube. Show that if $3c$ is not greater than $2b$ then C will come to the top of the vertical tube $\frac{(2b+c)\sqrt{3}}{\sqrt{2bg}}$ seconds after the motion begins.

VIII. PART III. SECOND PAPER.

[*Full marks will be given for about* two-thirds *of this paper. The acceleration due to gravity may be taken to be* 32 *foot-second units.*]

1. Prove that the tangent to a parabola at any point makes equal angles with the axis and the focal radius of the point.

The tangent at P to a parabola, whose focus is S, meets the directrix in Z and the latus rectum in D. Prove that the angle DSZ is twice the angle DPS.

2. The normal at P to an ellipse, whose centre is C, meets the major axis in G, and the ordinate of P meets the same axis in N; prove, with the usual notation, that

$$NG : CN = BC^2 : AC^2.$$

The ordinate NP is produced to meet the auxiliary circle in p, and the normal at P meets Cp produced in M; MR is drawn at right angles to the major axis; prove that

$$RG : CR = BC : AC.$$

3. Prove that the equation of any straight line can be expressed in the form

$$x \cos a + y \sin a = p.$$

Through the point, whose polar co-ordinates are r, θ, two straight lines are drawn, to make equal angles ϕ with the radius vector; prove that their equations are

$$x \sin (\phi \pm \theta) \mp y \cos (\phi \pm \theta) = r \sin \phi$$

the upper signs being taken together and the lower signs together.

4. Prove that the equation

$$x^2 + 3xy + y^2 - 7x - 3y + 1 = 0$$

represents two straight lines.

Find the angle between these lines, and prove that the co-ordinates of their point of intersection are -1 and 3.

5. Prove that, if the straight line

$$x \cos \alpha + y \sin \alpha = p$$

touches the circle

$$(x-h)^2 + (y-k)^2 = r^2,$$

then $$h \cos \alpha + k \sin \alpha - p = \pm r,$$

and explain the sign of the ambiguity.

A circle is described to pass through the origin, and to touch the lines $x = 1$, $x + y = 2$; prove that the radius r is a root of the equation

$$r^2(3 - 2\sqrt{2}) - 2r\sqrt{2} + 2 = 0.$$

6. Draw the curves $y^2 = 4(x+1)$ and $x^2 = 4(1-y)$.

The normal to the parabola $y^2 = 4ax$, at any point P, meets the axis in G, and GP is produced outwards to Q, so that $PQ = PG$; find the locus of Q.

7. C is the centre of an ellipse of semi-axes a and b; express in terms of the eccentric angle of a point P, the length of the semi-diameter conjugate to CP.

Prove that the acute angle between the two conjugate diameters cannot be less than $\sin^{-1} \dfrac{2ab}{a^2+b^2}$.

8. A uniform rod AB, of weight 2 lbs. and length 3 feet, can turn in a vertical plane about a hinge at A; a second uniform rod BC, also of weight 2 lbs., is hinged to the first at B, and the system is supported by a vertical cord attached to a point of AB distant 2 feet from A. Find the tension of the cord.

Find also the work done in very slowly raising the system from a position in which AB is horizontal, to one in which it makes an angle of 30° with the horizontal.

9. Prove that, when a system of forces, acting in one plane on a rigid body, keep it at rest, the sum of their moments about any point in the plane vanishes.

A uniform rectangular plate $ABCD$ is free to turn in a vertical plane about a hinge at A, and the plate is held in a position in which D is above and B is below the level of A by a horizontal force equal to the weight of the plate, applied at a point P of AB; prove that

$$AP = \tfrac{1}{2}(AD + AB \tan \alpha),$$

where α is the inclination of AD to the horizontal.

10. A shot of mass m is fired from a gun of mass M, placed on a smooth horizontal plane, and elevated at an angle α. Prove that, if the muzzle velocity of the shot is V, the velocity of recoil of the gun, at the instant when the shot leaves the muzzle, is

$$\frac{mV\cos\alpha}{\sqrt{\{M^2\cos^2\alpha+(M+m)^2\sin^2\alpha\}}}.$$

[All resistances are to be neglected.]

11. A particle is projected from a point A with a given velocity in a given direction; find the latus rectum of the parabolic path.

If B is any point of the path, C the middle point of AB, and D the point where the vertical through C cuts the path, prove that

$$CD=\tfrac{1}{2}gt^2,$$

where t is the time from A to D.

12. A point P describes a circle uniformly with angular velocity ω; find its acceleration.

Prove that, if O is the centre of the circle, and N is the foot of the perpendicular from P on a fixed diameter, the acceleration of N is $\omega^2\ ON$ towards O.

MATHEMATICAL EXAMINATION PAPERS

FOR ADMISSION INTO

Royal Military Academy, Woolwich,

NOVEMBER, 1901.

OBLIGATORY EXAMINATION.

I. ARITHMETIC. (1.)

(To vulgar fractions and decimals.)

[N.B.—*You are requested to put the* number *to each question, and to send up the* working *as well as the answers.*]

1. Add together 8 qrs. 5 bus. 1 pk., 2 qrs. 1 pk. 1 gal., 6 bus. 1 gal.

2. Divide £20. 0*s.* $1\frac{1}{2}$*d.* by 9.

3. Reduce 4040 pounds to tons, etc.

4. How many centimetres are there in 437 kilometres 5 metres?

5. Resolve into prime factors 2310.

6. Find the L.C.M. of 10, 15, 21, 35.

7. Express in its lowest terms $\frac{231}{330}$.

8. Add together $\frac{3}{8}$, $1\frac{3}{8}$, $3\frac{5}{16}$, $2\frac{1}{4}$.

9. From $3\frac{5}{6}$ subtract $2\frac{13}{18}$.

10. Multiply $2\frac{13}{21}$ by $8\frac{3}{11}$, giving the result in its lowest terms.

11. Divide 5 by $2\frac{1}{16}$.

12. Add together 5, 1·85, 2·9, 0·008.

13. Subtract 3·27 from 6·051.

14. Multiply 3·425 by 2·64.

15. Divide 803·6 by 560.

16. Express as a vulgar fraction in its lowest terms 0·96875.

17. Express as a decimal the ratio of 33 yds. to 2 furlongs.

18. Express ·0583 kilometres in metres and centimetres.

19. Find the cost of 235 articles at £4. 7*s.* 9*d.* each.

20. Find the value of 2 tons 9 cwt. 2 qrs. 8 lbs. at £4. 13*s.* 4*d.* per ton.

21. If the weight of 5 sovereigns is 41 grams, how many sovereigns will weigh 2 kilograms 50 grams?

22. Some earth is moved by 80 men in 13 hours: in how many hours would 100 men move $5\frac{1}{2}$ times the quantity?

23. How many litres are there in 32 per cent. of 5 hectolitres?

24. How much per cent. is 15*s.* 9*d.* out of £8. 15*s.*?

25. How many square feet are there in the floor of a room 19 feet long and 17 feet broad?

II. ARITHMETIC. (2.)

[*All working must be clearly shown.*]

1. A motor-car travels for the first 20 minutes at the rate of 6 kilometres 126 metres per hour, then, for 2 hours 20 minutes, at the rate of 30 kilometres per hour, and, afterwards, for 25 minutes, at the rate of 7 kilometres 152 metres per hour; find the total distance traversed by the car.

2. A kilogram is the weight of a cubic decimetre of water, and a decimetre is 3·937 inches. It being given that a cubic foot of water weighs a thousand ounces, express the kilogram, in pounds avoirdupois, correctly to two places of decimals.

3. An investor places 5670*l.* in Consols, $2\frac{3}{4}$ per cent., when they are at $94\frac{1}{2}$; find the income obtained. Also, find the amounts of money which he must invest in Canada Stock, 4 per cents., at 104, and in Victoria Stock, 3 per cents., at $97\frac{3}{4}$, in order that he may obtain, from each of these, the same income that he obtains from Consols.

4. A company borrows the sum of 1,658,775*l.* on the understanding that at the end of each year a portion of the principal is to be paid off, with interest at 4 per cent. per annum on the amount standing unpaid during that year. Prove that the debt can be cleared off in four years by an annual payment of 456,976*l.*

5. 17*s.* $11\frac{1}{2}$*d.* = 8 florins + 3 sixpences + 22 farthings.

$$= £\cdot 8.$$
$$+ £\cdot 075.$$
$$+ £\,(\cdot 022 + \cdot 001 \times \tfrac{22}{24})$$
$$= £\,(\cdot 897 + \cdot 001 \times \tfrac{22}{24})$$
$$= £\,\cdot 898 \text{ to the nearest thousandth.}$$

Check every step of the above working and hence deduce how any sum of English money may be written off at sight as a decimal of a pound sterling to the nearest thousandth.

Express 15*s*. 9¾*d*. as a decimal of a pound to the nearest thousandth.

6. Work out the following question, giving full reasons for each step in your solution :—

What is the least number that contains 8, 12 and 21 as measures?

III. PART I. First Paper.

[N.B.—*In questions on Geometry correct demonstrations, whether those of Euclid or not, will be accepted.*]

The circumference of a circle is 3·1416 *times the diameter.*

1. Resolve into 3 real factors

$$x^4 - 2.$$

Give as decimals to the nearest hundredth any surds that occur.

2. What is the sum of (*a*) the interior angles; (*b*) the exterior angles of a triangle? Deduce a formula for the magnitude of an angle of a regular polygon of *n* sides, and from this obtain the number of degrees in the angle of a regular pentagon.

3. In an acute-angled triangle *ABC*, *AO* is drawn perpendicular to *BC*. Denoting this perpendicular by *p*, and *OC* by *x*, equate the values of p^2 given by the two right-angled triangles and so determine *x* and thence *p* in terms of *a*, *b*, *c* (the lengths of the sides of the triangle). Deduce an expression for the area of a triangle in terms of the sides.

State, without proof, whether the expression applies for triangles generally.

4. The three angular points of a triangular plot of ground *ABC* are determined thus:—The plot lies in the angle between two paths *PQ* and *PR* which intersect at right angles. Starting from *P*, I can reach *A* by walking 15 yards along *PQ* and then 8 yards in a direction parallel to *PR*. The distances for *B* are 4 yards, 14 yards, and for *C* 4 yards, 3 yards, the starting point in each case being *P*, and the distances being stepped just as before. What is the area of the triangle *ABC*?

5. Show that ${}_4C_3 + {}_4C_2 = {}_5C_3$ by deriving the combinations of the 5 letters *a*, *b*, *c*, *d*, *e*, taken 3 at a time, from the sets of 3 which contain a particular letter and the sets of 3 which do not contain that letter.

In a similar manner, or otherwise, prove that

$${}_nC_r + {}_nC_{r-1} = {}_{n+1}C_r.$$

6. Draw a circle with a radius of 8 centimetres (or if you prefer it, 3 inches), and from a point *P* distant 16 centimetres (or 6 inches) from the centre *O* of the circle draw a tangent to the circle; measure and state its length.

If any secant *PAB* is drawn cutting the circle at *A* and *B*, show as well as you can by measurement and calculation, that the rectangle *PA* . *PB* is equal to the square on the tangent from *P* to the circle. (Two or three secants should be drawn and measured.)

7. Prove that if two equal triangles stand on the same base and on the same side of it they are between the same parallels.

A triangle is required having the same area as a given triangle but a greater base of given length. *Either*, show how to construct the required triangle and prove the method correct; *Or*, give the steps of the reasoning by which you arrive at a solution of the problem, and state the construction without further proof.

8. What is a locus?

Draw (1) the locus of a point at a constant distance of 2 ins. from a fixed point *A*; (2) the locus of a point equidistant from the points *B* and *C*, 3 ins. apart; (3) the locus of a point the product of whose distances from the fixed points *D* and *E* is 5 square inches, the distance from *D* to *E* being 4 ins. (In the last case find a few points and draw the curve free-hand.)

9. Give definitions of the sine and tangent of an angle.

Construct an angle of 15°, measure its sine and tangent and state your results; take 10 centimetres (or 5 inches if you prefer it) as the hypotenuse of the triangle from which sin 15° is obtained; also 10 cm. (or 5 inches) as base of triangle for tan 15°, and state to what degree of accuracy you consider your results reliable.

10. My compasses have legs 10 cm. long. I open them to an angle of 35° and describe a circle. What is the distance between the points of the compasses (to the nearest millimetre), and what is the area of the circle (to the nearest sq. cm.)?

IV. PART I. Second Paper.

[N.B.—*In questions in Geometry correct demonstrations, whether those of Euclid or not, will be accepted.*]

The circumference of a circle is 3·1416 *times the diameter.*

1. A workman is offered the option of being paid at the rate of 2*s.* a day or of receiving 6*d.* for the first day, 7*d.* the second day, 8*d.* the third day, and so on. For how many days must he work in order that he may receive the same amount whichever mode of payment he chooses?

2. Find what value must be assigned to a and b to make

$$3x + 4 = a(x - b)$$

for all values of x.

Find what values c, d and e must have to make

$$3x^2 + 10x + 3 = c(x - d)(x - e)$$

for all values of x.

3. If $a = x^k$, $b = x^l$, $c = x^m$, express, as a power of x, the fraction

$$\frac{a^p b^q}{c^r}.$$

Hence express the logarithm of this fraction, in terms of $\log a$, $\log b$ and $\log c$, the base being x.

4. Write down formulae giving the area of a circle, the volume of a right circular cylinder, and the volume of a sphere.

A boiler has the form of a right circular cylinder with two convex hemispherical ends; show that the area of its external surface is equal to the product of its greatest length and the circumference of the circular section of the cylinder.

5. Define *similar figures.* Prove that two triangles are similar if two sides of the one are proportional to two sides of the other and the included angles are equal.

A square $BCDE$ is described on the base BC of an acute-angled triangle ABC, and on the side opposite to A. If AD and AE cut BC in F and G respectively, prove that FG is the base of a square inscribed in the triangle ABC.

6. Prove the following construction for drawing a perpendicular on a given straight line from a given point P: take any two points A and B on the line, and with A and B as centres, describe circles passing through P; if these circles cut again in Q, then PQ is the perpendicular.

Criticise the following proof: "Let PQ cut the line in O, then in the triangles AOP and AOQ, AO is common, $AP = AQ$, and $\angle APO = \angle AQO$ (Euclid I., 5), therefore, $OP = OQ$ and $\angle AOP = \angle AOQ$, therefore, each of these angles is a right angle and OP is the required perpendicular."

7. Two straight lines AB and CD cross each other in O. How would you ascertain by measuring the lengths of the segments, whether the four points A, B, C, D lie on a circle or not?

In a diagram $OA = 8$ cm., $OB = 5$ cm., $OC = 4$ cm. and $OD = 14$ cm., and AB and CD intersect one another at an angle of $60°$. Determine, in centimetres, the distance from O at which the circle through the points A, B, C would cut the line CO produced.

8. Show (with proof) how to circumscribe a circle about a given triangle.

With your instruments make a triangle having sides 6, 8, 12 centimetres long and circumscribe a circle about it.

9. The latitude of London is 51° N., and the radius of the Earth 4000 miles. How far is London from the Equator measured along the Earth's surface, and how far from the Earth's axis?

10. In a triangle ABC the angles B and C are found to be $49° 30'$ and $70° 30'$ respectively, and the side a is found to be 4·375 inches. Find A, b and c as accurately as the Tables permit.

V. PART II. First Paper.

1. A man wanted to buy a horse. The owner said, "I'll give you the horse if you will buy the nails in his shoes; there are 20 nails, and you must pay a farthing for the first, a halfpenny for the second, and so on, twice as much for each nail as for the previous one." The buyer accepted. What had he to pay? An approximate answer will do.

2. From a quadrant AB of a circle an arc AP is marked off subtending an angle of $x°$ at the centre. A circle with centre A passes through P and cuts the chord AB in P'. Express AP' in terms of x.

Suppose the chord graduated so that every point P' corresponding to an integral value of x is marked x. How could you from a ruler graduated like this chord construct an angle of given magnitude?

3. Show that a set of values of x and y that satisfy at the same time

$$x^2+y^2-2=0, \quad \text{and} \quad 2x+y-3=0. \qquad \text{(1.)}$$

must satisfy

$$\left.\begin{array}{l}(x^2+y^2-2)+(2x+y-3)(x+2y-3)=0\\(x^2+y^2-2)-(2x+y-3)(x+2y-3)=0\end{array}\right\} \qquad \text{(2.)}$$

and

Find a set of values that satisfy equations (2) but do not satisfy equations (1).

4. From the tables find the values of $\tan 10x - 2\tan 9x + 1$ for the following values of x (in degrees):—

$$0,\ 1,\ 2,\ 3,\ 4,\ 5,\ 5{\cdot}8,\ 5{\cdot}9,\ 6,\ 7,\ 8,\ 9.$$

Draw on squared paper a curve showing how $\tan 10x - 2\tan 9x + 1$ varies with x when x is measured in degrees and lies between $0°$ and $9°$.

Find to the nearest tenth of a degree a value of x for which the given expression vanishes.

5. A vessel is formed by scooping a conical cavity out of a solid right circular cylinder whose height is 4 inches and the diameter of whose base is 5 inches. The base of the cone coincides with the base of the cylinder, and the vertex of the cone with the centre of the other circular end of the cylinder. Find in square inches to two decimal places the whole surface, internal and external, of the vessel.

6. A regular tetrahedron has edges 10 centimetres long. Find its volume to the nearest cubic centimetre.

Also find the angle between two faces to the nearest degree.

7. Find a positive value of x satisfying the equation $2^x = 7 \times 2^{\frac{1}{x}}$ approximately, so that the error in the value of x does not exceed one per cent.

8. If the base of a triangle is given in position and magnitude, and the ratio of its sides is also given, show that the vertex lies upon one or other of two circles, each of which divides the base internally and externally in the given ratio.

Show by means of a figure that the area of a triangle whose base is 12 cm., and whose sides are in the ratio of 7 : 5 cannot exceed 105 sq. cm.

9. With your instruments draw a triangle SPQ having $SP = 5$ cm., $SQ = 10$ cm., $PQ = 12$ cm. Consider S the focus of a conic of eccentricity 2, and P and Q two points on the curve, and construct the directrix corresponding to S.

Give one solution, stating your construction. State how many solutions there are.

10. If the normal at a point P of a conic meets the axis in G, show that the ratio $SG : SP$ is equal to the eccentricity, S being the focus.

VI. PART II. Second Paper.

The diagrams supplied are to be themselves used, and not pricked off into your book.

A body in a vacuum falls to the earth with an acceleration of 32 *foot-second units.*

1. State and explain the proposition known as the parallelogram of forces, pointing out how a force may be represented by a straight line.

A weight of 10 lbs. is supported by two strings inclined to the vertical at angles of 28° and 38°. Find the tension in each string.

2. In towing a boat along a river, is it easier to tow with a long rope or a short one, and why? (The weight of the rope is to be neglected.)

3. Explain the term *centre of gravity*.

Show how to find the centre of gravity of a quadrilateral lamina (*a*) by geometrical construction, (*b*) by experiment.

4. A uniform steelyard is 2 feet long and weighs 10 lbs. The pan weighs 35 lbs., and is hung from one end. The fulcrum is 3 inches distant from that end. A weight of 5 lbs. slides on the beam which is graduated to read to quarter-pounds. Find the distance between the graduations, and also the greatest weight which the balance can weigh.

5. A stone slab lies on a horizontal board, the coefficient of friction between them being 0·23. To what angle (to the nearest degree) can the board be tilted before the stone slips?

6. A string $ABCD$ is fastened to a fixed beam at A, passes under a movable pulley B, and over another movable pulley C, the power being applied at the end D. The pulleys B and C are held up by a string BEC, which passes over a fixed pulley E. Find the relation between the power and the weight which is attached to pulley B.

If the weight of the pulley C is 5 lbs. and that of B 4 lbs., what force at D will be required to balance a weight of 240 lbs.?

7. Define the terms ***uniform velocity, uniform acceleration.***

A point on the equator moves through 25,000 miles in 24 hours. Find its speed in feet per second (to the nearest integer).

8. A body starting from rest moves with a uniform acceleration of 32 foot-second units. Find how far it will move in the 1st, 2nd, 3rd, and 4th seconds respectively.

9. A train of 300 tons moving with a speed of 40 miles an hour is brought to rest by friction in 2 minutes (steam being shut off). Find the magnitude of the frictional force, which is assumed to be constant.

10. When a lift is at rest its floor is just strong enough to carry a load of 1 ton. If the lift is moving (*a*) with constant speed, (*b*) with an acceleration, can the floor support the same load?

11. Define the terms ***work, energy.***

A body of mass 1 pound is projected vertically upward with a velocity of 80 feet per sec. Find its kinetic energy and its potential energy after 1, 2·5, 5 sec., respectively, and show that the sum of its kinetic and potential energy is constant until it has struck the ground.

12. A particle moves with uniform speed in a circle of radius r feet, making n revolutions per second. Find an expression for the acceleration towards the centre.

VOLUNTARY MATHEMATICS.

VII. PART III. FIRST PAPER.

1. Explain how a curve may be represented by an equation.

Having given $y=\sqrt{a^2-x^2}$, where $a=2$ inches, find (by putting $x=a\sin\theta$ or otherwise) a number of sets of corresponding values of x and y; and trace the curve represented by the equation.

2. With mathematical instruments construct a triangle ABC in which $AB=7$ cm., $BC=10$ cm., and $CA=12$ cm. Draw the two conics which pass through A and have B and C as foci. Determine their eccentricities.

3. PNP' is a chord of a parabola perpendicular to the axis AN, and the normal at P cuts the axis in G; show that NG is of constant length.

QGQ' is a chord parallel to PNP' of the circle whose diameter is PP'; show that the locus of Q for different positions of PP' is another parabola.

4. Show that the locus of the point of intersection of any tangent to an ellipse and a perpendicular line through a focus is a circle.

An ellipse has a given focus and touches three given straight lines; find the centre of the ellipse and the points of contact of the given tangents.

5. Show that the four straight lines given by the equations $6x^2-5xy-6y^2=0$ and $6x^2-5xy-6y^2+x+5y-1=0$ lie along the sides of a square.

6. Pairs of points P, Q are taken on a parabola such that their distances from the axis of the parabola differ by a constant. Show that the chord PQ touches another parabola.

7. Find the equations of the asymptotes and axes of a conic given by the equation $48x^2-16xy-15y^2+112x-14y+56=0$.

8. Show that, if three forces keep a body in equilibrium, their lines of action will meet in a point or be parallel.

A smooth uniform hemisphere is in equilibrium with its curved surface resting on two parallel horizontal fixed rods which are not in the same horizontal plane. Show that the plane face of the hemisphere must be horizontal.

9. Show that a system of forces acting in one plane is equivalent to a single force or a single couple.

AB, CD are two equal uniform rods which can turn freely about the two points A and C respectively, the line AC being horizontal. To B and D are fastened the ends of a weightless string which passes through a ring R whose weight is half that of either of the rods AB, CD. Show that, if in the position of equilibrium AB and BR make angles α and β respectively with the vertical, then $\tan\beta = 3\tan\alpha$.

10. Equal weights each of mass m are attached to the end of a fine string which passes over two fixed smooth pulleys and under a smooth movable pulley, also of mass m; the pulleys being so arranged that all the parts of the string which are not in contact with the pulleys are vertical. Find the upward acceleration of the movable pulley and the tension of the string.

VIII. PART III. Second Paper.

[*The acceleration due to gravity may be taken to be 32 foot-second units.*]

1. Show that a linear equation represents a straight line. Show on a diagram the lines $x=3, \quad x+y=2, \quad x=y.$

Find the co-ordinates of their intersections and the area of the triangle contained by them.

2. Prove that all parabolas are similar curves.

If G be the foot of the normal at a point P of the parabola, R the middle point of SG (S being the focus) and X the foot of the directrix, show that the difference between the squares on RX and RP is equal to the square on the semi-latus rectum.

3. On a given diagram two branches of a hyperbola were shown. It was required to draw the axes and the director circle. State your construction.

4. Find the angle at which the circles

$$(x-\alpha)^2+(y-\beta)^2-r^2=0$$
$$(x-\alpha')^2+(y-\beta')^2-r'^2=0$$

intersect.

Find the general equation of the circles which cut the circle

$$x^2+y^2-2x+2y-2=0$$

at right angles at the point (1, 1).

5. Prove that if two normals to a parabola are at right angles they intersect on a fixed parabola. Find the dimensions and position of this parabola, and show it in a figure.

6. Investigate the equation of the normal at a given point of the conic

$$ax^2+by^2=1.$$

Find the co-ordinates of all the points on this conic such that the normals drawn through them pass through one or other of the extremities of the minor axis.

7. Determine the centre of any system of parallel forces acting in one plane.

Parallel forces of 10, 18, 14 and 6 lbs. weight act at the angular points, taken in order, of a square. Find the position of the centre of parallel forces when the forces 10 and 18 act in one direction and the forces 14 and 6 in the opposite direction.

8. Two posts have been put up, and a hammock is to be slung symmetrically between them to carry a man of given weight just clear of the ground. Find the tension of the rope for a given inclination. (The dimensions of the man are small compared to the length of the rope.)

Does the liability to pull down the posts depend on the height at which the rope is fixed to the posts?

9. Investigate the motion of two unequal weights connected by a cord which passes over a smooth fixed pulley.

A heavy string hangs at rest over a small smooth fixed peg. Show that if $\frac{198}{199}$ths of the string on one side be cut off, the pressure on the peg is instantaneously reduced to $\frac{1}{100}$th of its previous amount.

10. A projectile in vacuo has an initial velocity V. It is required to hit the top of a tower h feet high which stands on the horizontal plane through the point of projection and subtends an angle α at that point. Show how to determine the angle of elevation, *either* by calculation *or* by geometrical construction.

MATHEMATICAL EXAMINATION PAPERS

FOR ADMISSION INTO

Royal Military Academy, Woolwich,

JULY, 1902.

OBLIGATORY EXAMINATION.

I. PART I. FIRST PAPER.

[N.B.—*In Geometry the demonstrations and sequence of propositions need not be those of Euclid.*]

1. On a certain map a road 1320 yds. long is represented by a length of $18\frac{3}{4}$ inches. Determine the scale of the map. What area on the map would represent an area of $\frac{1}{10}$ sq. mile?

2. A litre of water weighs a kilogram, a litre of another liquid weighs 1·340 kilograms. A mixture of the two weighs 1·270 kilograms per litre. Determine the volume of each in a litre of the mixture.

3. Four points A, B, C, D are taken in order at random on the circumference of a circle and joined so as to give the quadrilateral $ABCD$. Show that, whatever their position, the sum of the opposite angles is constant.

If at A, B, C, D tangents KL, LM, MN, NK are drawn, prove that the sum of one pair of opposite sides of the quadrilateral thus formed is equal to the sum of the other pair.

4. Solve the equations

$$\text{(i)}\quad 3x+4+2(x-7)=3(2x+4);$$

$$\text{(ii)}\quad \frac{3{\cdot}2x+2{\cdot}1}{17}=\frac{3{\cdot}4x+5{\cdot}6}{14};$$ giving x to one decimal place, *i.e.* to the *nearest* tenth.

5. A regular pentagon is inscribed within a circle of radius r. Show that its perimeter is $10r\sin 36°$, and its area is $5r^2\sin 36°\cos 36°$, and find its perimeter and area as nearly as the tables allow when $r=5$ inches.

6. Find, either by Algebra or by Geometry, a point C in the line AB (a inches long), so that the sum of the squares on AC and CB shall be equal to a given square (side b inches), and find the length of AC to the nearest hundredth of an inch when $a=4$, $b=3$.

7. Calculate the following by logarithms, and show how you would roughly check your results :—

(i) pr^n, where $p=93{\cdot}75$, $r=1{\cdot}03$, $n=4$;

(ii) $\frac{4}{3}\pi r^3$, where $\pi=\frac{355}{113}$, $r=5{\cdot}875$.

8. Draw figures to show that

(i) $(a+b)^2=a^2+2ab+b^2$;

(ii) $(a-b)^2=a^2-2ab+b^2$;

(iii) $(a+b)(a-b)=a^2-b^2$.

Figures alone—drawn on a large scale—will suffice if the various parts are clearly indicated.

9. A line AB is bisected at O and divided unequally at C. Prove, as briefly as you can, that

$$AC^2+CB^2=2AO^2+2OC^2.$$

State, or illustrate graphically, what changes would take place in the magnitude of AC^2+CB^2 if the point C move along the line from A to B.

10. Expand $(1+a_1)(1+a_2)(1+a_3)(1+a_4)$, and show how $(1-a)^4$ may be derived from the result.

If $a=0{\cdot}0003$, find, correct to 4 decimal places (*i.e.* to the nearest ten-thousandth) the value of $(1-a)^4$.

II. PART I. SECOND PAPER.

[N.B.—*In Geometry the demonstrations and sequence of propositions need not be those of Euclid.*]

1. A bicycle wheel 28 inches in diameter makes 132 revolutions per minute. Express its speed in miles per hour. ($\pi = \frac{22}{7}$.)

A cyclist with a 28-inch wheel says that he can find his speed in miles per hour by simply counting the number of "clicks" made by his cyclometer in 5 seconds, each click indicating a revolution of the wheel. Test this assertion.

2. Express as a single fraction in its lowest terms

$$\frac{1}{x-1}+\frac{1}{x-2}+\frac{1}{x+1}+\frac{1}{x+2}.$$

3. If $y=ax+b$, what numerical values must be given to a and b so that $y=-11$ when $x=-2$, and $y=+1$ when $x=+2$?

4. What is a parallelogram, and what special kinds are there?

Prove that a parallelogram may be obtained by each of the following constructions:—

(*a*) Set off equal lengths AB and CD on two parallel lines and join corresponding ends by the lines AC and BD.

(*b*) Join the extremities of any two lines AB and CD that bisect each other at any angle.

What other conditions must be imposed in (*b*) to get each of the special kinds of parallelogram?

5. The angles α and β are acute, $\sin\alpha=\frac{4}{5}$ and $\sin\beta=\frac{5}{13}$. Calculate the value of $\sin(\alpha+\beta)$ and of $\alpha+\beta$.

Construct a triangle ABC in which AD the perpendicular from A on BC is 6 cm. long, and the angles DAB and DAC are the angles α and β. Measure the angle BAC, and compare it with the value already found for $\alpha+\beta$.

6. Construct an angle whose sine is 0·72, determine the angle and its tangent by measurement, and compare these results with those obtained from the tables. Take 10 cm. as the hypotenuse of the right-angled triangle you first construct.

7. The sides of a rectangular piece of paper are measured both in centimetres and in inches, and the length of a metre in inches is determined from each set of measurements. Find, to 2 decimal places, the average of the two results if the length of one side of the rectangle is 16 inches or 40·6 cm., and that of the other 12·75 inches or 32·5 cm.

8. Determine *with your instruments* whether the following construction will give a regular octagon inscribed in a square $ABCD$ (taking each side of the square to be 4 inches long):—"Draw the diagonals AC and BD intersecting at X. With A, B, C, D as centres, and AX as radius, describe arcs cutting the sides of the square. The points of intersection are the angular points of the octagon."

Indicate the nature of the tests you apply. Verify your conclusion by calculation or general reasoning.

9. Show that if an object of height h at distance d from the observer subtends a small angle of A degrees at his position, then roughly $h = \frac{Ad}{57{\cdot}3}$.

Use this to find the height of a tower which subtends an angle of 9° at a point 170 yards away.

10. I want a ready means of finding approximately 0·866 of any number up to 10. I select a point O at the corner of the squared paper where two thicker lines cross, and find a second point P by going 10 inches to the right and then 8·66 inches up (or 5 to the right and 4·33 up) and join O to P. The two thick lines passing through O are scaled off in inches, OX to the right, OY up. Explain clearly why the distance from OX of any point in OP is 0·866 of its distance from OY.

Read off from the scales, and mark on the appropriate places on the paper, 0·866 of 3, 0·866 of 6·5, 0·866 of 4·8, and $\frac{1}{0{\cdot}866}$ of 5.

III. PART I. THIRD PAPER.

[N.B.—*In Geometry the demonstrations and sequence of propositions need not be those of Euclid.*]

1. The hypotenuse and one side of a right-angled triangle are 5·61 and 3·24 inches respectively. Find the length of the other side to the nearest hundredth of an inch.

2. Divide x^2+px+q by $x-a$ till a remainder is obtained independent of x, and show how the remainder might be at once derived from the dividend.

Give the remainder when $x^3-3x^2-18x+40$ is divided by $x-2$.

3. Prove that any triangle is equal to half a rectangle with the same base and height.

Construct a triangle ABC in which $AB=9$ cm., $BC=10$ cm., $CA=8$ cm., and by measurement find the length of the perpendicular from A to BC, then use this result for finding the area of the triangle.

4. Suppose that BXC is an angle whose tangent is 2. Suppose that with X as centre and any radius XC, a semicircle BCA is described, and CD is drawn perpendicular to XB. Show either by construction and measurement or by general reasoning that CD is the side of a square inscribed within the semicircle.

5. Imagine a chord AB to revolve about a fixed point X within a circle, and then suppose the chord itself fixed and the point X to move along it from A to B. Discuss the changes that take place in the length of the chord and in the magnitude of the rectangle $AX.XB$.

6. My house roof has a boundary enclosing a horizontal area of 1700 square feet, and the annual rainfall is 18 inches (*i.e.* the rain if collected where it fell would cover the ground to a depth of 18 inches). If I depend for my water supply on the rain falling on my roof, how many gallons may I use per day? (A gallon is 277 cubic inches.)

7. Find values of x that satisfy the equations

$$\text{(i) } 4x^2+7x-2=0,$$
$$\text{(ii) } 4x^2+2x-1=0,$$

expressing as decimals to the *nearest* hundredth any surds that occur. Test your results by substituting in the equations the values found for x.

8. Given that the curved figure in a diagram is equal in area to the rectangles shown, find, by measuring the rectangles, the area of the figure in square centimetres. [The breadth of each rectangle is 1·3 centimetres, and the lengths are 3·1, 5·6, 6·8, 7·2, 7·1, 6·2, 5, 4·1, and 2·2 centimetres.]

The quadrilateral figure $ABCD$ represents a field, and is drawn on the scale of 1 inch to a chain (22 yards). Find, by measurement, as accurately as you can, the length of a path running direct from A to C, and the total area of the field. Indicate, on the diagram, the length of each line measured. [In the diagram $AB=4{\cdot}2$, $BC=2\frac{1}{4}$, $CD=4\frac{3}{4}$, $DA=1{\cdot}8$, and $BD=4{\cdot}7$ inches.]

9. Under what conditions are two triangles similar? Mention all possible cases arising from consideration of sides and angles of the two triangles.

You are told that two sides and the included angle of one triangle are 2 in., $2\frac{1}{2}$ in. and 60°, and that the three sides of anothei aie 4 in., 4·3 in., and 5 in. Determine with your instruments whether the triangles are similar, indicating the nature of your test, and verify your conclusion by calculation.

10. Write down two formulæ which can be used for determining the angles of a triangle when the three sides are known.

Either, determine the greatest angle of a triangle whose sides are 7 yds., 8 yds., 9 yds., and confirm your result by measurement of a figure drawn to scale;

Or, find the greatest angle of a triangle whose sides are 6·28 in., 5·32 in., and 7·64 in.

IV. PART II. First Paper.

1. A cubic foot of water weighs $62\frac{1}{2}$ lbs. Find the size of a hollow cube and of a hollow sphere, each of which will hold a gallon (10 pounds) of water. Give the edge of the cube and the diameter of the sphere approximately in inches.

2. Find the total amount of money paid in Income Tax in 9 years on an annual income of £960, if Income Tax is at the rate of 8*d.* in the pound the first year, and rises 2*d.* in the pound each year up to 1*s.* 8*d.* and then remains steady.

3. What are the acute angles whose cosines are equal to $\frac{1}{2}$ and $\frac{1}{2}\sqrt{3}$ respectively?

The hour hand of a church clock is 30 centimetres long. How many centimetres does its extremity rise (*a*) between 6 and 8 o'clock; (*b*) between 8 and 9 o'clock?

4. The temperature taken every two hours one day showed:

Midnight	$41^{\circ}\cdot0$	2 p.m.	$51^{\circ}\cdot2$
2 a.m.	$40\cdot8$	4 p.m.	$53\cdot0$
4 a.m.	$40\cdot7$	6 p.m.	$46\cdot5$
6 a.m.	$39\cdot5$	8 p.m.	$46\cdot3$
8 a.m.	$40\cdot8$	10 p.m.	$46\cdot7$
10 a.m.	$44\cdot5$	Midnight	$47\cdot4$
Noon	$48\cdot0$		

Draw a curve to show the variation of temperature throughout the day, and estimate the temperature at 3 p.m.

5. A man surveying a road goes first 7 chains in a direction 63° east of south, then 8·3 chains 80° east of south, then 12 chains 46° east of north, and then 5·7 chains 16° west of north. How far is the final point east of the starting point, and how far north? And in what direction and at what distance from the starting point? You may answer graphically or by calculation.

6. **Write down or obtain expressions for the volume of (i) a pyramid on a square base; (ii) a truncated pyramid on a square base.**

Find the volume of air contained in a street lamp with a flat top and bottom, the four slanting faces of the lamp being equal trapeziums, whose sides are 6, $9\frac{1}{2}$, 12 and $9\frac{1}{2}$ inches long.

7. Find, in terms of the length of an edge, the distance of a vertex of a regular octahedron from the plane through the four neighbouring vertices.

Give the distance, to the nearest centimetre, when the edge of the solid measures 10 cm.

8. A straight line L **is perpendicular to a plane** P. **Show that the projection of** L **on any plane** Q **is perpendicular to the trace of** P **on** Q, ***i.e.*** **to the intersection of** P **and** Q.

9. With your instruments find a number of points on a conic of eccentricity 0·7 with the focus 1·5 inches from the directrix, and draw the curve freehand.

10. S **is the focus of a conic,** P **and** Q **are two points on the curve on the same side of the directrix, and** PQ **produced meets the directrix in** Z. **Prove that** SZ **bisects the exterior angle between** SP **and** SQ.

Given the focus, directrix and one point P **on the curve, show how to find any number of points on the curve.**

V. PART II. SECOND PAPER.

[Ten *questions and no more are to be answered.*]

A body in a vacuum falls to the earth with an acceleration of 32 foot-second units.

1. What do you understand by the expression *the resultant* of two forces?

Find the resultant of the two forces, 30 lbs. acting in the direction PA and 10 lbs. acting in the direction BP, when the angle $APB = 51°$.

2. State the condition that must be satisfied if any number of forces acting on a particle hold it at rest.

Four forces act outwardly at a joint of a frame. One force of 6 tons is horizontal; two others of 3 tons and 11 tons respectively make angles of 60° and 171° (measured in the same direction) with the horizontal force. Find graphically the amount of the remaining force when there is equilibrium.

3. Define the moment of a force with respect to a point.

A force of 10 lbs. acts tangentially at the circumference of a pulley 2 feet in diameter. What force must be applied to a rope coiled round the shaft to which the pulley is keyed (or fixed) to prevent the pulley from turning? The diameter of the shaft is 3 inches.

4. A beam carrying a load of 5 tons at 3 feet from the left-hand support, rests freely with its ends upon two supports 20 feet apart.

Calculate the pressure on each of the supports. The beam itself is of uniform section, and weighs 2 tons.

5. A cubic centimetre of concrete weighs 2·5 grams. Find the crushing load in kilograms per square centimetre on the base of a wall of the concrete with parallel vertical faces 6 metres high and one metre wide.

6. A train weighs 200 tons. Find what force must be exerted on the drawbar connecting the train to the engine to draw the train up an incline of 1 in 50 (that is 1 foot vertically for 50 feet horizontally), the resistance to traction on the level being 12 pounds-weight per ton. Calculate the work done (in foot-tons) if the incline is 5 miles long.

7. Find the energy stored in a train weighing 250 tons and travelling at 60 miles per hour. How must energy must be added to the train to increase its speed to 65 miles per hour?

8. *ABDEC* is a uniform steel plate half an inch thick, made up of a square *AKEC* of 20″ side and a trapezium *EKBD* with the side *BD* parallel to *KE* and 5″ in length and the side *KB* 10″ in length. Find its mass and the distances of its centre of gravity from *AB* and *AC* respectively. A cubic inch of iron weighs ¼ lb.

9. A stone is dropped from rest from the top of a tower 200 feet high. Neglecting the frictional resistances of the air, find how long it takes to reach the ground, and calculate its velocity just before it touches the ground. Plot a curve roughly showing the velocity of the stone at various heights.

10. Assume the telegraph poles of a railway to be 50 yards apart. A train is observed to pass a post every four seconds. Calculate its speed in miles per hour. After an interval of five seconds it is observed to pass a post every three seconds. What is then the speed? Calculate in feet per second per second the acceleration (supposed uniform) during the interval of 5 seconds.

11. An ordinary block and tackle (the system of pulleys in which there is one continuous rope) has two pulleys in the lower block and two in the upper block. If the system were frictionless, what force would have to be exerted to lift a load of 400 pounds? If the efficiency

of the system is actually 0·4, *i.e.* if a given force will lift only 0·4 times as much as if the system were frictionless, what force would be required to lift the load of 400 pounds?

12. Rain is falling vertically, speed 5 miles an hour, on a man walking at 4 miles an hour. At what angle to the vertical must he hold his umbrella to make the rain drops hit the top at right angles? Give the angle to the nearest degree, determining it in any way you please.

VI. PART III. FIRST PAPER.

[N.B.—Eight *questions, and not more, are to be answered.*]

1. Explain how the position of a point in a plane is fixed by co-ordinates.

Find an expression for the distance between (x_1, y_1) and (x_2, y_2), and apply it to find to the nearest hundredth the distance between (3, 7) and (5, 3·4).

2. Draw the locus represented by $y = \frac{3}{5}x + 1$, taking an inch as unit. Find its inclination to the x axis to the nearest degree.

3. Find the co-ordinates of the centre of mass of m_1 at (x_1, y_1) and m_2 at (x_2, y_2).

Where is the centre of mass of two equal masses at (1·32, 4·7) and (3·14, 7·08)? Show the two points and the centre of mass on the squared paper (unit 1 inch).

4. Show that the equation of a circle is of the second degree.

Find the co-ordinates of the intersections of $x + y = 11$ and $x^2 + y^2 = 61$.

5. Having given a focus and directrix, ascertain for what values of the eccentricity the conic will be entirely on one side of the directrix.

Where do the two vertices lie when the focus is an inch from the directrix and the eccentricity is 2? Answer this by a freehand sketch, marking on it the distances of the vertices from the focus.

6. Solve $2x^2 + y^2 - 2xy - 2y + 1 = 0$, first as an equation for y, and then as an equation for x. Hence, or otherwise, ascertain for what values of x, y is real, and for what values of y, x is real. Finally, trace the curve represented by this equation.

7. Find the equation of a tangent to a conic at any point of the curve.

Which of the straight lines $x = 0$, $x = 1$, $x = 2$, is a tangent to

$$2x^2 + 3xy + y^2 + x + y + 1 = 0?$$

Ascertain whether the others cut the curve in real points.

8. Prove that the locus of the middle points of a series of parallel chords of an ellipse is a straight line.

Prove that the tangents at the points where this locus cuts the curve are parallel to the series of chords.

9. A train is travelling northwards at 60 miles an hour and the wind is blowing from the south-west at 20 miles an hour. Show, in a diagram, the direction of the trail of smoke of the engine.

10. The resistance to a train weighing 100 tons travelling at the rate of 60 miles an hour is 10 pounds-weight per ton. Find the rate of working of the engine.

The train consumes half a ton of coal per hour, and the burning of a pound of coal produces 11 million foot-pounds of energy. What proportion of the energy is usefully employed by the engine?

11. If the rotation of the earth were stopped, would a mass suspended from a spring balance stretch the spring moie or less? If the earth rotated 10 times as fast as at present, could you jump higher or not so high? Give your reasons.

VII. PART III. SECOND PAPER.

[Eight *questions and not more are to be answered.*]

The acceleration due to gravity may be taken to be 32 *foot-second units.*

1. Explain how an equation may represent a curve, and show that an equation of the first degree represents a straight line.

2. Show that the straight line joining the points (20, -15) and (-24, 18) passes through the origin, and find the co-ordinates of the point half-way between them.

3. Find the equation of the circle which has its centre at (2, 1) and passes through the origin.

Find the length of its radius to the nearest hundredth.

4. Find the co-ordinates of the middle point of the part of the line $y=3x$ that lies inside the circle

$$x^2+y^2+3x-6y-2=0.$$

5. Trace the curve $y=2x-x^2$, taking an inch as unit.

Find, approximately, the co-ordinates of the point where it cuts the line $y+4=0$.

6. Ascertain whether $x^2+2xy-3y^2+2x-3y=0$ is an ellipse, a parabola, or a hyperbola.

Trace the curve.

7. Prove that the tangent at a point P on an ellipse bisects the exterior angle between SP and $S'P$, S and S' being the foci.

State the corresponding results for the hyperbola and parabola, illustrating your statements by figures.

8. From any point P on a hyperbola a parallel to each asymptote is drawn to cut the other. These points of intersection being M and N, prove that the rectangle $PM \cdot PN$ is constant.

Draw a hyperbola with perpendicular asymptotes, for which the value of the rectangle $PM \cdot PN$ is one square inch.

9. A string *ASPQRUB* passes over 2 pulleys *S* and *U* and carries 2 weights at its ends, that at *A* being 100 grams. At *P* and *Q* also weights are hung and at *R* a force acts downwards inclined to the vertical at an angle of 19°.

If the system comes to equilibrium with the parts of the strings *SP*, *PQ*, *QR*, *RU* inclined to the vertical at angles of 34°, 63°, 298°, 325°, find the amount of the weight at *PQ* and *U* to the magnitude of the force at *R*.

10. A ring *P* is strung on the smooth wire shown by a curve in a figure and is let go from a position 1 inch vertically above the lower end of the wire. Will it fly off the wire? Give your reasons.

11. A ditch is bridged over by a light plank *AB*, supported at its two ends. A man of weight *W* stands at any point *P* of the plank.

(i) Find what portions of his weight are supported by the reactions at the two ends *A* and *B* respectively.

(ii) Take any point *Q* between *P* and *B*, and prove that the moment about *Q* of the forces acting on either one of the portions *AQ* or *QB* is

$$W.\frac{AP.QB}{AB}.$$

(iii) If this moment measures the tendency of the plank to break at *Q*, prove that the plank will be most likely to break in the middle when the man is standing there.

MATHEMATICAL EXAMINATION PAPERS

FOR ADMISSION INTO

Royal Military Academy, Woolwich,

NOVEMBER, 1902.

OBLIGATORY EXAMINATION.

I. PART I. FIRST PAPER.

[N.B.—*In Geometry the demonstrations and sequence of propositions need not be Euclid's.*]

1. In 1900-1901 the national expenditure was £183,592,264. Of this the Army cost £91,710,000, the Navy £29,520,000, and the Civil Services £23,500,000. Express each of these as a percentage of the whole, to the nearest integer.

2. Make geometrically an equilateral triangle 4 square inches in area, and measure its side as correctly as you can. State your construction.

Also calculate the length of a side and compare your results.

3. If $v^2/r = g/289$, calculate v, having given that

$$r = 4000, \quad g = \frac{32 \cdot 2}{5280}.$$

Also show that the value of

$$\frac{2\pi r}{v \times 60 \times 60}$$

where $\pi = 3 \cdot 1416$, is approximately 24.

4. Draw two straight lines *OB* and *OC* at right angles, and *OA* between them making 39° with *OB*. With centre *O* and radius 10 cm. draw a circle cutting *OB* in *Q* and *OA* in *P*. From *P* let fall perpendiculars *PS* on *OB* and *PR* on *OC*. At *Q* draw a tangent *QT* cutting *OA* in *T*. Measure *PR*, *PS*, and *QT* to the nearest millimetre, and write down their lengths. Hence find sin 39°, cos 39°, tan 39°, and compare with the values given in the tables.

5. Prove in any way you please that approximate values of $2^{\frac{1}{3}}$ and $2^{\frac{7}{12}}$ are $\frac{5}{4}$ and $\frac{3}{2}$ respectively. Find in each case whether the correct or the approximate value is the greater, and find also in which case the approximate value is most nearly equal to the correct value.

6. A railway is 8·1 metres broad and 4372 kilometres long. Find to the nearest square kilometre, the area it occupies.

7. *AB* is a fixed straight line. Find the locus of a point *P* which moves so that $MP^2 = AM \cdot MB$ where *PM* is the perpendicular drawn from *P* on *AB*.

8. A set square has its hypotenuse 12 inches long and the shorter side 4 inches. The hypotenuse slides along a scale which is held fixed, and an arrowhead on the hypotenuse is placed in succession against marks at intervals of 0·15 of an inch on the scale. In each position a line is ruled along the longer side of the set square. How far apart are these lines?

If an error of 0·01 of an inch was made in placing the arrowhead, what error in the position of the line would result?

9. The volume of a tree-trunk is taken to be equal to the length multiplied by the square of a quarter of the girth. How much per cent. would this be in error if the trunk was a circular cylinder? ($\pi = 3 \cdot 1416$.)

10. Show from first principles that $a^3 \times a^5 = a^8$.

Calculate $\log \frac{347 \times 0{\cdot}231}{7{\cdot}62}$.

If $\log a^{\frac{2}{3}} = 0{\cdot}5126$, find a.

II. PART I. SECOND PAPER.

[N.B.—*In Geometry the demonstrations and sequence of propositions need not be Euclid's.*]

1. The return railway ticket from a suburb to London costs 2*s.* 2*d.*, and the annual ticket £12 5*s.* 0d. How many days in the year must a man travel to save money by taking an annual ticket instead of paying for each day?

2. Find the ratio of x to y from the equation $2x^2 - 9xy + 10y^2 = 0$.

3. With your instruments construct a square $ABCD$ whose side is 2 inches, and produce AB to F cutting off $BF = 2{\cdot}83$ inches. By a geometrical construction apply to BF a rectangle equal in area to the square $ABCD$, stating your construction.

4. The price of a standard troy ounce of silver on 1st January in each of the ten years (1891–1900) was (in pence)

45, 40, 36, 29, 30, 31, 28, 27, 27, 28.

Draw a curve showing its value approximately at any time during these ten years.

5. A gunner at P wishes to know the distance of a point Q held by the enemy. Suppose R a third point and that he knows $PR = 4325$ yards, $QPR = 70°$, $QRP = 64°$, and calculate the distance PQ for him. Check your result by drawing a figure in which PR is represented by 4·325 inches, and measuring PQ.

6. The basin of the Severn is 4350 square miles in area, and the yearly rainfall 29 inches. Half of the rain that falls is again evaporated and the rest flows to the sea. Find, to the nearest million gallons, how much water flows to the sea per hour. (A cubic foot of water is 6·23 gallons.)

7. Given the base and vertical angle of a triangle, state what is the locus of the vertex, and quote a proposition in geometry which establishes your conclusion. (The number of the proposition is not required.)

Draw, with your instruments, two straight lines at right angles, AB and AC, one 2 inches and the other 3 inches long. Find a point at which AB and AC both subtend angles of 120°. State your construction.

8. State and prove the relation between the three sides of an acute-angled triangle and the projection of one side on another.

Deduce an expression for an angle of the triangle in terms of the sides.

9. For a certain book it costs a publisher £100 to prepare the type and 2*s*. to print each copy. Find an expression for the total cost in pounds of x copies.

Also make a diagram on the scale of 1 inch to 1000 copies and 1 inch to £100, to show the total cost of any number of copies up to 5000. Read off the cost of 2500 copies, and the number of copies costing £525.

10. A very long strip of paper is rolled on a cylinder of wood 1 inch in diameter, and the total diameter of the roll is 2 inches. Find what would be the total diameter of the new roll if the same strip of paper were coiled on a cylinder of wood 2 inches in diameter.

III. PART I. THIRD PAPER.

[N.B.—*In Geometry the demonstrations and sequence of propositions need not be Euclid's.*]

1. In 1901 the National Debt was £673,608,200. How much would be required to pay interest on this for a year at $2\frac{3}{4}$ per cent.?

2. What is the greatest and least number of obtuse angles that a quadrilateral (other than a rectangle) can have? How is the result modified if the quadrilateral is known to be cyclic (*i.e.* if a circle can be drawn about it)?

In a quadrilateral $ABCD$, $AB=6$ in., $CD=8$ in., $AD=BC=3$ in., and AB is parallel to CD. If the quadrilateral is freely jointed together, could you distort it so as to give it 3 acute angles? Justify your statement. (No internal angle is to exceed 180°.)

3. Knowing the number of pounds in a cubic inch of a substance, you can find the number of kilograms in a cubic cm. by multiplying by $0 \cdot 4536 \times (2 \cdot 54)^{-3}$. Express this multiplier as a decimal to 3 places.

If steel weighs 488 lbs. per cubic foot, how many kilograms per cubic centimetre does it weigh?

4. I know that the time of the half-swing of a pendulum l feet long is either $2\pi\sqrt{\frac{l}{32 \cdot 2}}$ seconds or $\pi\sqrt{\frac{l}{32 \cdot 2}}$ seconds, where $\pi = 3 \cdot 1416$; I am uncertain which of these is correct, but I know that the half-swing of a pendulum a yard long takes about a second. Find from these data, with as little work as possible, which is the correct expression.

5. Calculate in decimal form the various terms of the expansion of $(1+\frac{3}{100})^5$ by the binomial theorem.

Find the compound interest on £100 for 5 years at 3 per cent.

6. An English standard of deal planks contains 120 planks 12 feet long, 9 inches wide, and 3 inches thick. How many cubic feet of wood does it contain?

7. If AD be drawn bisecting the vertical angle A of a triangle ABC and meeting the base in D, prove that

$$BA/AC = BD/DC.$$

If the lengths of the sides opposite the angles A, B, C, be a, b, c respectively, express BD and DC in terms of a, b, and c.

8. A surveyor finds two points A and B in a hillside to be 3 chains 43 links apart, and finds the line AB to be inclined at $17° \, 30'$ to the horizontal. On his plans these points must be shown at their horizontal distances apart. What is this, to the nearest link? (1 chain is 100 links.)

Check your calculation by measurement from a drawing made to a scale of 2 inches to 1 chain.

9. A workman is to be paid 1*s.* for his first day's work, 1*s.* 1*d.* for the second day, 1*s.* 2*d.* for the third day, and so on increasing at the rate of a penny each day that he works. Find how much more he earns in the second week than in the first, if he works six days in the week.

10. A haystack is in the form of a prism 20 feet long, with vertical plane ends of the shape shown in the annexed figure, and the following

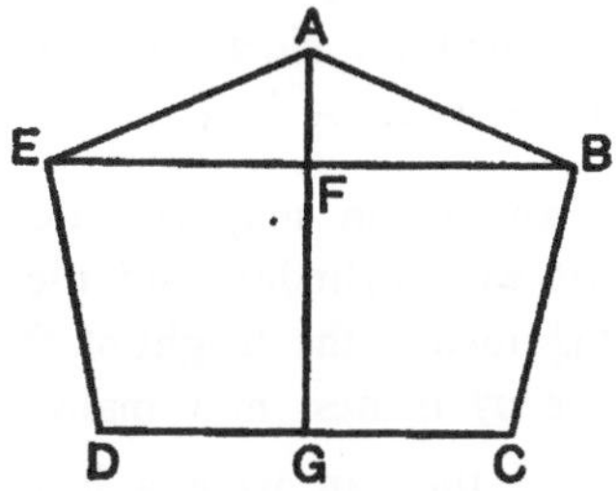

(Not drawn to scale.)

measurements are given:—

$$EB = 12 \text{ feet}, \quad DC = 10 \text{ feet}, \quad AF = 2\tfrac{1}{2} \text{ feet}, \quad FG = 6 \text{ feet}.$$

Find the number of cubic feet of hay in the stack.

IV. PART II. First Paper.

1. The surface of a sphere of radius r inches is $4\pi r^2$ square inches, where $\pi = 3{\cdot}14159$. If π is taken, instead, to be equal to $\frac{22}{7}$, find, roughly in square miles, the difference that this will make in the calculated area of the earth's surface, the earth being a sphere of radius 4000 miles.

2. Calculate to five decimal places (*i.e.* to the nearest hundred-thousandth) the value of $x^2 - 280x + 95$ when x is $0{\cdot}339697$ and when x is $0{\cdot}339698$.

Give to six decimal places two values of x for which the given expression vanishes, explaining how you find them.

3. Determine with your instruments whether $\sin 2A$ is greater than, equal to, or less than twice $\sin A$, taking A as 35°. Draw a large figure, and indicate the nature of your test.

Write down an expression for $\sin 2A$ in terms of $\sin A$ and $\cos A$, and verify from the Tables when A is 35°.

4. The capacity of an oil-can may be calculated approximately by regarding the lower part as a cylinder, and the upper part as a cone. If the total height is $14\frac{1}{2}$ inches, the height of the cylindrical portion 10 inches, and circumference 27 inches, how many gallons will it hold?

If the circumference of the can were a inches, what would be the circumference of another of the same shape but of double the capacity?

$$[\pi = \tfrac{22}{7},\ 1 \text{ gall.} = 277 \text{ cub. in.}]$$

5. $ABCD$ is a parallelogram; $AB = 2{\cdot}5$ inches, $BC = 4$ inches, and $\underline{|ABC} = 65^\circ$. Calculate the area of the parallelogram and the lengths of the diagonals.

Test your results by drawing a figure and measuring.

6. A straight line DE is drawn parallel to AC a side of a $\triangle ABC$ and cuts the other sides BA, BC in D and E respectively. Write down expressions for the areas of BDE, BDC, and BAC in terms of BE, BC, BD, BA, and the angle ABC. Hence express the ratio BDE/BAC in terms of BE and BC.

7. *OX*, *OY*, *OZ* are edges of a rectangular box. Putting x, y, z for *OX*, *OY*, *OZ* respectively, express the diagonals of the faces and also the diagonal *OA* of the box in terms of x, y, z.

If $x = 6$ cm., $y = 8$ cm., $z = 5$ cm., find the length of *OA* and the angles which *OA* makes with the plane *XY* and with the edge *OX*.

The results may be obtained either by calculation or by measurement from figures in which the true lengths are shown.

8. (i) On the squared paper choose two of the thicker lines meeting at right angles, *OX* to the right, *OY* up. Scale off *OX* to show time (1 inch for 1 hour), *OY* to show distance (1 inch for 10 miles). From *O* draw a line diagonally which will indicate distance travelled in a specified time at the rate of 12 miles an hour. Explain how to use it, and illustrate by reading off the distance travelled in 4·4 hours and the time required to travel 36 miles.

(ii) Hence, or otherwise, work the following:

Two cyclists start at noon, one from London at the rate of 12 miles an hour, the other from Dover 72 miles distant, at 10 miles an hour. At what time and at what distance from London will they meet, and at what times will they be 20 miles apart?

9. Take as directrix of a conic a straight line *XY* and as focus a point *Z* 3 inches from *XY*. One point on the curve is 3 inches from *XY* and 1 inch from *Z*. Find a number of points on the curve and draw it.

10. Show (with proof) how to draw a tangent to a conic from a point outside the curve.

V. PART II. SECOND PAPER.

[Ten *questions, and not more, are to be answered. The acceleration of a body falling freely in vacuo near the Earth's surface may be taken as* 32 *feet per second per second.*]

1. Find the resultant of a force of 52 lbs. and two forces of 95 and 75 lbs. making with the first force angles of 76° and 230° respectively (the angles being measured in the same direction). Write down the magnitude of the resultant and the angle it makes with the force of 52 pounds-weight.

2. AC is a horizontal bar 32 inches long fixed by a smooth axis at A, and freely jointed at C to another bar, BC, 67·5 inches long, which is fixed by a smooth axis at the end B, this point being vertically under A. A load, W, of 600 lbs. being suspended from C, find the force exerted in each bar, the weights of the bars being negligible. Your solution may be graphical or by calculation.

3. AB is a uniform bar 8 feet long, having a mass of 150 lbs. supported in a horizontal position on two vertical props at A and B. How near an end may a load of 200 lbs. be placed if the pressure on a prop may not exceed 250 lbs.?

4. Two strings are so connected that when one is pulled down 3 feet the other rises 1 foot. If you hang a weight of 5 lbs. on the former, what pull must you exert on the latter to balance it? Give your reasons.

5. In a system of 4 pulleys consisting of a fixed block which contains 2 of them, and a moveable block containing the others, the same rope passes continuously round all the pulleys. From the moveable block is suspended a basket containing a man, the weight of man and basket being W. If another man standing on the ground pulls the free end of the rope, what force must he exert to raise the man and basket?

If the free end of the rope is pulled by the man in the basket, what force must he exert for the same purpose?

6. *AC* is a straight line divided at *D* so that $AD = 1{\cdot}7$ and *DC* is $1{\cdot}5$ inches. *AB*, of length $2{\cdot}2$ inches, and *DE*, of length $3{\cdot}2$ inches, are drawn perpendicular to *AC* on opposite sides. In any way you please find the distance of the centre of gravity of the quadrilateral *ABCE* from *AC*.

7. If the block of a steam hammer weighs half a hundredweight and has moved through a distance of 8 inches under the action of a uniform force of 10,000 pounds-weight, with what momentum is it then moving? (Name your units.)

8. What force, in pounds-weight, will produce an acceleration of 8 feet per second per second in a mass of 12 lbs.?

A man of weight 150 lbs. stands on a platform; what pressure does he exert on it if the platform moves vertically with an acceleration of 2 feet per second per second (1) upwards, (2) downwards?

9. Define kinetic energy and work, naming any units in which they are measured.

If a particle whose mass is 4 ounces has a velocity of 1000 feet per second, through what distance will it overcome a force of 360 pounds-weight before coming to rest?

10. The Moon has a mass of $1{\cdot}7 \times 10^{24}$ pounds, and describes a circular orbit about the Earth of radius $1{\cdot}26 \times 10^{9}$ feet in 27 days 8 hours. Calculate the Moon's acceleration and the force that produces it. State your units.

11. If a shot of 140 lbs. is fired from a 5-ton gun with a muzzle velocity of 1600 feet per second, and the recoil of the gun is taken up by causing it to ascend an inclined plane, through what *vertical* height will the gun rise? (Neglect the weight of the powder charge.)

VI. PART III. First Paper.

[*A body in a vacuum falls with an acceleration of 32 foot-second units.*]

1. Draw on squared paper the line

$$2x+0{\cdot}7y=2{\cdot}5 \text{ inches.}$$

Find to one decimal place the length of the perpendicular to this line from the origin.

2. Find, to one place of decimals, the length of the tangent from the point (5, 6) to the circle $x^2+y^2-4x+4y=0$.

You may do this graphically on squared paper if you like.

3. Solve the equation $2x^2+2y^2+5xy+1=3x+3y$ for y. Give the values of y when $x=0{\cdot}1$, $0{\cdot}2$, $0{\cdot}5$.

Draw the locus represented by the equation.

4. Show that if all the chords of an ellipse parallel to the minor axis are lengthened in the same properly chosen ratio, their middle points being held fixed, their extremities will lie on a circle.

In what ratio is the area enclosed by the curve increased by this change?

5. A diagram shows a hyperbola of which O is the centre. Suppose its equation to be $ax^2+2hxy+by^2=1$ (unit 1 inch). What can you tell about the coefficients from the fact that OY is an asymptote? What are the co-ordinates of its intersections with OX, and what can you tell from this about the coefficients? Complete the determination of the coefficients in any way you like, giving their values to two decimal places.

[The diagram showed the hyperbola $x^2+4xy=3$.]

6. A slab, 5 lb. in weight, rests on a horizontal plane. A horizontal pull of 1 pound-weight would just move it. To what angle (to the nearest degree) can the plane be tilted before the slab slips?

7. The oldest proposition in Statics is that of Archimedes, that "Weights are in equilibrium when they are inversely proportional to their distances from the point of support." How would you ascertain whether this is true? Assuming its truth, explain how it is applied in the common balance.

8. A simple pendulum consisting of a weight hanging by a light string from a fixed point swings through an angle of 20° on each side of the vertical. Find the greatest velocity of the weight, the length of the string being 5 feet.

VII. PART III. SECOND PAPER.

[*In a vacuum a body falls with an acceleration of* 32 *foot-second units.*]

1. Place the points (1, 2) (−3, 2) (2, −1) (−2, −3) on squared paper taking any rectangular axes and the inch as unit.

State the co-ordinates of the mid-points P and Q of the lines joining the first two points and the last two, and find the abscissa of the point where PQ cuts the axis of x.

2. Find the equation of the circle which passes through the two points $(a, 0)$ $(-a, 0)$ and whose radius is $\sqrt{a^2+b^2}$.

Regarding a as constant and b as variable, prove that the polar of a given point (α, β) with respect to any circle which passes through two fixed points passes through another fixed point.

3. Show that the tangent at any point of a hyperbola makes equal angles with the focal distances.

Having given the foci F and F' and a point P in the conic, show how to draw the tangent and normal at P.

4. Plot the points (0, 0) (1, 0·5) (2, 2) (3, 4·5) (4, 8) (5, 12·5), taking 1 inch as unit. Prove that a parabola with the y-axis for axis of the curve can be drawn through these points. Find its equation and draw it.

5. A and B are two fixed points 4 inches apart, and a point P moves so that the sum of its distances from A and B is 5 inches. Take as axes of co-ordinates the line AB and the line bisecting AB perpendicularly, and find the equation of the locus of P.

Find where the curve cuts each axis.

6. The position of a point P on a straight line is given by

$$x = t^2 - 2t + 3,$$

where t denotes the number of seconds after a fixed instant of time, and x the distance in feet of P from a fixed point O on the line. What are the positions of P at the fixed instant, and at intervals of 1 and 3 seconds after the fixed instant?

What space does P traverse between time 3 seconds and time 6 seconds, and between times 3 seconds and 3·001 seconds, and what is the average speed for each of these intervals? What is the actual speed at time 3 seconds?

7. A string with one end fixed runs horizontally to a pulley, passes round the pulley, and carries a weight of 56 lbs. at the other end. Find the thrust of the pulley on its axle, neglecting the weight of the pulley and the cord. Give its magnitude to two significant figures and state its direction.

8. Supposing the resistance of the air on a falling raindrop to be proportional to the square of its speed, show that for a certain speed the acceleration is zero, so that a drop that has attained this speed will continue to fall with constant speed.

If the resistance of the air lessens the downward acceleration of the drop by 2 foot-second units when the speed is 10 feet per second, what is the constant speed?

MATHEMATICAL EXAMINATION PAPERS

FOR ADMISSION INTO

Royal Military Academy, Woolwich,

JUNE, 1903.

OBLIGATORY EXAMINATION.

I. PART I. FIRST PAPER.

[N.B.—*In Geometry the demonstrations and sequence need not be Euclid's.*]

1. The best railway run in Britain is from Darlington to York, 44 miles in 43 minutes; the best in America is from Camden to Atlantic City, $55\frac{1}{2}$ miles in 50 minutes; and the best in France, Paris to Arras, 193 kilometres in 117 minutes. Arrange these runs in order of speed, taking a kilometre as 0·62 of a mile.

2. Taking the figure on the Diagram as a map on a scale of $\frac{1}{1000000}$, measure AB and calculate the real distance between A and B in kilometres (or, if you prefer, in miles).

[The diagram showed 3 points A, B, C, placed so that $AB = 2{\cdot}62$ in., $BC = 5{\cdot}35$ in., and $CA = 3{\cdot}35$ in.]

3. A boy has a sheet of metal in the shape of a triangle whose sides are 9, 11, and 16 centimetres, and wants to cut from it as great a circle as possible. Find by geometrical construction the diameter of the greatest circle he can get, and write down its length in centimetres.

4. Find the air space in a hall whose dimensions are—length 60 ft., width 38 ft., height from ground to eaves 21 ft., height to ridge 35 ft.

5. The real distance of C from A (*see* Diagram of question 2) is 9·08 kilometres and from B 13·47 kilometres, and the angle ABC is $37° \ 10'$. Calculate the angle BAC and check your result by measuring it.

6. The earth describes its path round the sun in 31556925·51 seconds. Express this interval in days, to the nearest ten-thousandth.

Find the difference between the above period and the average length of the year, estimated on the basis that of 400 years 303 are of 365 days, and the remainder leap years of 366.

7. Draw a circle 2·6 inches in radius and take a point O 4·4 inches from its centre. Consider a straight line OPQ passing through O, and cutting the circle in P and Q. Draw this line in four positions, including one position in which P and Q coincide, and by measuring OP and OQ in each position prove that the rectangle $OP \cdot OQ$ is constant.

Find an equation connecting the radius (r) of the earth, the height (h) of a mountain, and the distance (d) of the sea horizon as seen from the mountain top.

8. Find from the tables the logarithms of 400, 401, 402, 403, 404, 405, 406. Find the excess of each over the logarithm of 400.

On squared paper represent the increments of the logarithms corresponding to the addition of 1, 2, 3, 4, 5, 6 to the number 400. Show how to find, with the help of your Diagram, log. 403·2, log. 403·5, log. 403·3. Compare your result with that given by the table of differences.

9. A dam has to be built of material g times as heavy as water; and its thickness in feet T at any depth d feet is calculated from the formula $T = g^{-\frac{1}{2}} d$. When the material is 2·3 times as heavy as water, what must the thickness be at a depth of 8 feet?

10. In London at mid-winter at noon when the sun's altitude is known to be 28° a factory chimney casts a shadow 190 ft. long. Calculate the height of the chimney.

II. PART I. SECOND PAPER.

[N.B.—*In Geometry the demonstrations and sequence need not be Euclid's.*]

1. A rectangular field is 18·3 chains long and 14·5 chains wide. Give the area in acres to the nearest acre. 10 square chains = 1 acre.

2. The magnetic needle at Greenwich points at present to the west of north. The following were the values in recent years of the angle of deviation of the needle from the north :—

1898.	1899.	1900.	1901.	1902.
16° 40′	16° 34′	16° 29′	16° 23′	16° 17′

Find the average rate of decrease of the deviation annually. What will be the probable value of the deviation in 1905?

3. The earthwall along the bank of a river is 12 metres broad at the base, and 6 metres broad at the top, and the sides slant at 40° to the horizontal. It is proposed to dig away the outer part of the wall and throw it on the top, and so form a wall whose section is an isosceles triangle, the slant of the sides being 40° as before. Determine, with your instruments, the height of the wall in its new shape. State your construction and the reasons for it.

4. Find the values of x and y which satisfy the equations $y=x+2$, $y=bx+3$. Drawing the graphs of the equations on squared paper, indicate the values of x and y which satisfy both equations (i) when $b=0{\cdot}7$; (ii) when $b=0{\cdot}8$; (iii) when $b=0{\cdot}9$. Is it possible to find values of x and y which satisfy the two equations when $b=1$?

5. Find, from the definitions, a formula which will give $\cos\theta$ when $\tan\theta$ is known.

Taking $\theta = 34° \; 43'$, find its tangent from the table; then find its cosine from your formula, and compare the result with that given in the tables.

6. The components of gunpowder are:—nitre 75 per cent., charcoal 15 per cent., and sulphur 10 per cent. How many grams of each (to the nearest gram) would you need to make a pound (454 grams) of gunpowder?

7. Make a triangle *ABC*, having $AB = 7$ in., $\hat{A} = 50°$, $\hat{B} = 40°$. On *BC* take *E* 1 in. from *B* and *F* 1 in. from *C*. Draw *ED* parallel to *BA*, and *FD* parallel to *CA*, meeting in *D*. By measurement and calculation find the ratios $\frac{BC}{EF}$, $\frac{BA}{ED}$, and $\frac{CA}{FD}$ as decimals to two places.

State a property that supplies a test of the accuracy of your work.

8. A manufacturer has priced certain lathes; the largest sells at £175 10*s*., and the smallest at £40. He wishes to increase his prices so that the largest will sell at £200 and the smallest at £50. Assuming that the new price (*P*) and the old price (*Q*) are connected by the relation $Q = a + bP$, find the values of *a* and *b* and, to the nearest pound, the new prices of lathes originally valued at £150, at £125 10*s*., and at £78. Use squared paper or algebra as you please.

9. A water-tube boiler has 350 tubes of 2·5 inches internal diameter, and the length of each tube is 8 ft. Find the total heating surface (*i.e.* interior surface) of the tubes in square feet.

[Area of curved surface of cylinder $= 2\pi rl$, where r is the radius, l the length, and π is 3·14.]

10. From a survey the dimensions of a quadrilateral field *ABCD* are $AB = 98$ yds., $AC = 138$ yds., $AD = 147$ yds., $BAC = 34°\ 15'$, and $CAD = 22°\ 30'$. By means of the expression $\frac{1}{2} bc \sin A$ for the area of a triangle calculate the area of the field in square yards.

III. PART I. THIRD PAPER.

[N.B.—*In Geometry the demonstrations and sequence need not be Euclid's.*]

1. The area of the section of a boring is 1325 square feet, and the excavating machine is driven forward 4 ft. a day. How many cubic yards of earth are excavated in a day?

2. Show that two parallelograms on the same base and between the same parallels are equal in area.

As an illustration, draw on the squared paper a rectangle with base 2 ins. long and height 3 ins., and on the same base a parallelogram that has the same height and has a side 5 ins. long, and count the number of small squares in each.

3. The height H of a mountain in feet is given by

$$H = 49{,}000\left(\frac{R-r}{R+r}\right)\left(1+\frac{T+t}{900}\right),$$

where R, r are the observed heights of the barometer in inches, at the bottom and top of the mountain, and T, t are the observed temperatures at the bottom and top. The following observations were made at the bottom and top of a certain mountain:—

	Barometer.	Thermometer.
At bottom -	29·60 in.	67°
At top -	25·35 in.	32°

Find the height of the mountain, and roughly check your result by any method which occurs to you.

4. The height h of the eye above the sea, and the distance d of the horizon, are connected by the equation $h^2 + 2hR = d^2$, where R is the earth's radius. Taking $R = 3960$ miles, find, in feet, the height above the sea at which the distance of the horizon is 5 miles.

5. The corner-post C of a property was fixed as being 87·6 chains from a tree, and in the direction *S.* 56° 50′ *E.* This post having now been moved to a point C' 25 chains due north of C, the distance and direction of C' from the tree must be determined. Find them by calculation.

6. In 1902 there were 462,593 children attending London Board Schools. The amount levied in school rates was £2,339,540. How much per child does this amount to?

7. AB and AC are two straight lines at right angles. A point X moves so that the difference of the perpendiculars from it to AB and AC is constant. Find the locus of X by plotting on squared paper, and justify your result by geometrical reasoning.

8. Taking the circumference of a circle as 3·1416 times the diameter, test, by drawing, for a circle 5 cm. in diameter, the truth of the following construction:—Draw a diameter CA; draw AB perpendicular to CA and 3 times as long; from the centre O draw a radius OD, making 30° with OC; and let fall DP perpendicular to CA. Then PB is equal to the circumference.

9. The population of Ireland was 4,458,775 at the census of 1901, and is decreasing, so that in x years from the census it falls to $(0·9477)^x$ of the population at the census. Estimate the population now, $2\frac{1}{4}$ years after the census.

10. The distance between two places shown on a map is the horizontal distance, but a surveyor has often to measure up or down a slope. The distance measured in this way is greater than the horizontal distance, and to reduce it to the horizontal distance he multiplies by a number depending on the slope. Make out a table of multipliers for slopes of 5°, 10°, 15°, 20°, 25°.

IV. PART II. FIRST PAPER.

1. A right circular cone was measured in such a way that the diameter (d) of the base is known to be between 16·2 and 16·3 inches, and the height (h) between 27·5 and 27·6 inches. Find the volume of the cone, using (1) the lesser, (2) the greater, dimensions. If you give your answer to seven significant figures how many are useless?

The volume of the cone is $\frac{\pi h d^2}{12}$ where $\pi = 3{\cdot}1416$.

2. P and Q are two forts on the same side of a straight entrenchment. A base line XY of 1000 yds. is measured along the entrenchment and the following angles are observed:—

$$Y\hat{X}P = 95^\circ,\ X\hat{Y}P = 43^\circ;\ X\hat{Y}Q = 105^\circ,\ Q\hat{X}Y = 27^\circ.$$

Find the distance between the forts, and check your result by drawing a plan to a scale of 6 in. to the mile. You may find useful the formula $\tan\frac{B-C}{2} = \frac{b-c}{b+c}\cot\frac{A}{2}$, or the formula $a = (b+c)\cos\phi$ where ϕ is given by $(b+c)\sin\phi = 2\sqrt{bc}\cos\frac{A}{2}$.

3. A rough sketch of the railway from A to B is made up as follows:

AC, 13 chains straight N. 70° E.

CD, an arc of a circle of 15 chains radius, subtending 20° at the centre.

DE, 26·5 chains straight East.

EF, an arc of a circle of 13 chains radius, subtending 30° at the centre.

FB, 26·9 chains straight S. 60° E.

The mark "N. 70° E." against a line means that its direction lies between North and East and makes 70° with the North. Calculate the length of the railway from A to B.

4. Expand $\left(1-\frac{a}{D}\right)^{-1}$ in powers of a to the term in a^3.

When a balance is used in air of density a and a body of density D is balanced in it by brass weights of density d weighing m grams, the real weight of the body in grams is

$$m\frac{1-\frac{a}{d}}{1-\frac{a}{D}}.$$

Expand this expression to the first power of a, and find how much the weight of the body differs from that of the brass weights when $m=1000$, $a=0{\cdot}0012$, $d=8$, $D=0{\cdot}5$.

5. A man has 100 sq. ft. of iron plate, and wishes to make with this a rectangular cistern, on a square base, without a lid. If x feet be the length of a side of the base, determine the capacity of the cistern in cubic feet.

Also, by plotting on squared paper the value of this capacity, determine the value of x which will give a maximum capacity, and the number of cubic feet in this maximum capacity. Use scales of 1 in. to the ft. and 1 in. to 10 cubic ft.

6. When are two figures said to be similar?

OA, OB, OC are three straight lines drawn from O. On OA measure $OP=5$ cm., $OP'=7{\cdot}5$ cm.; on OB measure $OQ=3$ cm., $OQ'=4{\cdot}5$ cm.; on OC measure $OR=4{\cdot}5$ cm., $OR'=6{\cdot}75$ cm. Prove, by measurement, that triangles PQR, $P'Q'R'$ are similar, and justify the result by geometrical reasoning. Find by any method the ratios area $PQR : OP^2$ and area $P'Q'R' : OP'^2$. Are the ratios the same? Give reasons.

7. A tetrahedron, vertex A, base BCD has edge $AD=BC=5$ cm., edge $BD=AC=4$ cm., edge $AB=CD=3$ cm. Imagine the three faces ABC, ABD, ACD, turned outwards about BC, DB, CD respectively till they are in the plane of BCD. Draw accurately the figure so obtained.

How many degrees are there in the sum of the three plane angles at each corner of a tetrahedron of which each edge is equal to the opposite edge? Prove it.

8. Through how many miles an hour does London move in consequence of the rotation of the Earth? Take the Earth as a sphere of radius 3960 miles. London is in latitude 51° 30′.

9. Calculate the eccentricity of the conic with focus *S* and directrix *MN* that passes through *T*.

Find out if this conic would pass through *V*. State your method.

The distances of *S*, *T*, and *V* from *MN* are 1·34, 2·21 and 2·85 ins., and the distances of *T* and *V* from *S* are 3·23 and 4·45 ins.

10. Draw the tangent and normal at *T* to the conic of Question 9. State your method.

V. PART II. SECOND PAPER.

[*The diagrams supplied are to be themselves used, and not pricked off into your book.*

Take the acceleration of a falling body as 32 foot-second units.]

1. Three forces 90 pounds weight in direction BA, 50 pounds weight in direction CA, and 80 pounds weight in direction AD act on the point A which is free to move in any direction. $BAC = 145°$, $BAD = 132°$. Find what force must be added to the group to keep the point at rest. Write down the magnitude of the force and the angle it makes with the force AB.

2. Two parallel forces whose magnitudes are respectively 10 and 17 pounds weight, act in the same plane and in opposite directions, in lines 10 feet apart. Find the magnitude and position of their resultant, stating the position with reference to the larger force.

3. A bar AC carrying a load of 56 pounds weight at C is fixed to the ground at A, and a prop BC is applied fixed to the ground at B and hinged at C. Find the respective loads on AC and CB. The angles of inclination of AC and BC are 30° and 70° respectively.

4. A force of 3 pounds weight acts at the end C of a straight horizontal lever ABC, making with the horizontal an angle whose tangent is $\frac{4}{3}$. The length of the lever is 11 ft. and the distance of the fulcrum B from A is 6 ft. What force must be applied vertically at A to balance it? Use graphical or arithmetical methods as you prefer.

5. Masses 3 lbs., 5 lbs. and 9 lbs. are placed at the vertices A, B and C of the plane triangle ABC. $BC = 4''\cdot 2$, $AC = 3''\cdot 4$ and $AB = 2''$. Find the mass centre of the system.

6. A body is suspended by a fine string at a distance 20 feet vertically above the ground. The string is cut. Find the time taken by the body to reach the ground.

7. A mass of 500 lbs. moving in a straight line is observed to change its speed from 60 miles per hour to 50 miles per hour in 10 seconds. What force must have acted to produce the change?

8. A mass weighing 300 lbs. is moving in a straight line with a uniform speed of 20 feet per second. Calculate its kinetic energy, and its momentum. State the units in which you express your answers.

9. A particle is moving in a circular path, 5 feet radius at a speed of 30 feet per second. If the mass of the particle is 10 pounds, what force must act to constrain the motion? Make a sketch, showing clearly how the force must act supposing the particle is constrained in its path by a string attached to it and to the centre of the path. What would be the motion of the particle if the string were cut?

10. By the principle of work, or otherwise, find the relation between the power and the weight in the case of an ordinary screw jack in which the screw is of half-inch pitch and the radius at which the power is applied is 2 feet.

VI. PART III. FIRST PAPER.

[*Take the acceleration of a falling body as 32 foot-second units.*]

1. Graphically or by calculation find the coordinates of centre of gravity of three equal masses at (0, 1), (1, 2) and (3, 0).

2. Calculate the coordinates of the intersections of

$$x^2+y^2-4x+2y+1=0 \text{ and } 3x+4y-1\cdot 5=0,$$

and the length of the line between the points of intersection.

Check your results by a graphical solution.

3. Find the equation of a parabola shown in a diagram referred to its axis, and the tangent at its vertex as axes of coordinates.

4. What is meant by a pair of conjugate diameters of an ellipse? State their principal properties.

5. Show that if $x=t+\frac{1}{t}$ and $y=t-\frac{1}{t}$, the locus of the point (x, y) when t varies is a hyperbola.

6. Show that the roots of the equation

$$y^4-4y^3+5y^2-9=0$$

are the ordinates of the points of intersection of the circle $x^2+y^2=9$ with the parabola $(y-1)^2=x+1$.

Find two of the roots by drawing the curves on squared paper.

7. A mass of 370 pounds is pulled up a rough slope of 1 in 20 (so that going 20 inches along the slope gives a rise of 1 inch) by a rope parallel to the plane, whose tension is 1800 poundals. If the resistance due to friction is 1100 poundals, find the velocity acquired by the mass in 10 seconds. Give the velocity to the nearest foot per second.

8. A light triangular frame *ABC*, with smooth pin joints at *A*, *B*, and *C*, is supported at *A* and *C* with *AC* horizontal, and *AB*, *CB* inclined at 20° and 47° to the horizontal. If a weight of 1000 lbs. is hung at *B*, find graphically the force acting in each of the three bars.

9. A pile weighing 4 cwt. is being driven into the ground by a weight of 3 cwt. let fall on it from a height of 13 feet. If each blow drives the pile $\frac{3}{4}$ of an inch, show that resistance of the earth to penetration is between 13 and 14 tons weight. (Assume this resistance to be a constant force, and that after the blow the weight and the pile move on together till stopped by this force.)

VII. PART III. SECOND PAPER.

[*Take the acceleration of a falling body as 32 foot-second units.*]

1. On the squared paper draw a quadrilateral with its corners at (3, 5), (1, 4), (0, 3), (3, 1), the unit being 1 inch.

Find the area of the quadrilateral in any way you please, stating your method.

2. Find the equation of the circle with radius 1·5 and centre at (1·5, 2). Show that it touches the line $x=0$.

Show the figure on squared paper, unit 1 inch.

3. Given the position of the locus and directrix of a parabola and a given point P, draw the tangents from P to the parabola. State your construction.

4. P is any point on the curve $\left(\frac{x}{4}\right)^2 - \left(\frac{y}{3}\right)^2 = 1$, Q is the point (5, 0), and PM the perpendicular from P on the line $x=3·2$. Calculate the lengths PQ and PM, and show that their ratio is independent of the position of P on the curve.

5. With 2 inches as unit, plot the points (0, 1), (1, 0·33), (2, 0·09), (3, 0), (4, 0·33), (3, 1), (2, 1·24), (1, 1·33), (0·5, 1·30), (0·5, 0·53), (1·5, 1·31), (1·5, 0·19), (2·5, 1·14), (2·5, 0·02), (3·5, 0·80), (3·5, 0·03); and draw a smooth curve through them.

Draw also the locus of the middle points of chords parallel to the y axis.

6. Bullets are dropped at successive intervals of $\frac{1}{10}$ of a second. How far has one fallen before the next is let go, and how far apart are these two when the first has fallen 10 feet?

7. In a balance the arms are 12 cm. long, the mass of the beam is 200 grams and the mass of each scalepan 65 grams. The centre of gravity of the beam is 2 mm. below the point of support. Each pan is hung so that the centre of gravity of the pan, whether loaded or not, is always vertically below the point of support of the pan, and the points of support of the beam and pans are in a straight line. Find the angular deflection of the beam when the load in one pan is 50 grams and that in the other 50·1 grams.

8. The governor of a steam engine, *ABDC*, is a frame of jointed rods each 6 inches long, carrying at *B* and *C* two equal heavy balls, and at *D* a sleeve which slides on the vertical rod *AE*. When the frame is turning about *AE* at 80 revolutions a minute, what is the inclination of *BA* to the vertical? All masses but those of the balls at *B* and *C* may be neglected.

9. The figure shows the position at any time of a body moving along a straight line, the ordinate giving the distance the body has traversed at any time. The slope of the curve at any point represents the speed of the body at the corresponding time (unit 10 feet per second). Find the slope at the four points *D*, *E*, *F*, *G*, and so find the speeds at these times; and draw another diagram to show the speed at any time.

[The *slope* of the curve at any point is the tangent of the angle made with the time-axis by the tangent to the curve at that point. You may draw tangents by laying a straight-edge against the curve.]

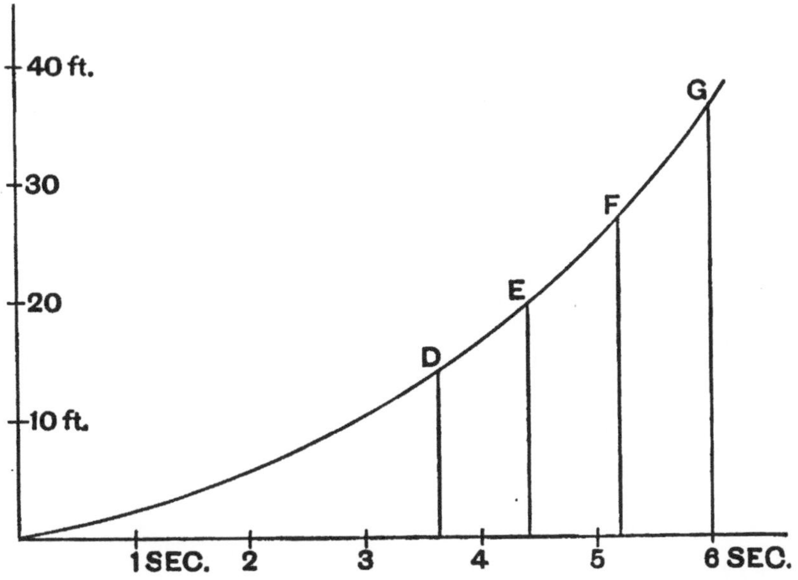

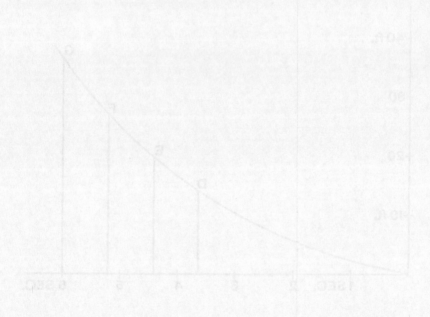

MATHEMATICAL EXAMINATION PAPERS

FOR ADMISSION INTO

Royal Military Academy, Woolwich,

NOVEMBER, 1903.

OBLIGATORY EXAMINATION.

I. PART I. FIRST PAPER.

[N.B.—*In Geometry the demonstrations and sequence need not be Euclid's.*]

1. A cubic foot of water weighs 62·3 pounds, and 41 cubic feet of teak weigh a ton. Is teak heavier or lighter than water, bulk for bulk?

2. A quadrilateral $ABCD$ has $AB = AD$ and $\angle ABC = \angle ADC$. Prove that $BC = DC$.

It is required to reproduce an exact copy of this quadrilateral. How many measurements must be made? Indicate what parts (*i.e.* sides and angles) might be measured for the purpose, giving the least number in each case.

3. In a triangle ABC, AD is drawn perpendicular to BC. Prove that

$$BD^2 - DC^2 = BA^2 - AC^2.$$

Given $BC = 66$ feet, $CA = 32$ feet, $AB = 40$ feet, find BD and DC, and hence calculate the angles of the triangle through the cosines of the angles at B and C.

4. Four people wish to be conveyed as quickly as possible to a place 5 miles distant. They can walk 4 miles an hour and have a motor car that can go 18 miles an hour but only holds two besides the driver. Two set off on foot; the car takes the other two a certain distance, sets them down to continue on foot and then returns to bring the first two. At what point must the car set down those it took first in order that the four may arrive at the same instant?

Find also the time taken to do the journey.

5. In the diagram erect perpendiculars at intervals of 1 centimetre along the straight line *AB*, beginning at *A*. Join the points at which successive perpendiculars meet the curve. Estimate approximately the area of the given figure by assuming it to be equal to the polygon so made.

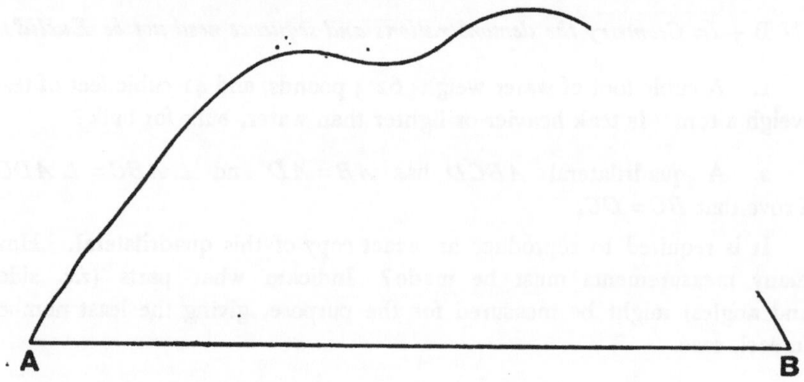

6. I wish to form a rough estimate of the acreage of a rectangular field by stepping along two adjacent sides. If my steps average 33 inches, and the two sides are 57 and 159 steps respectively, show that the field contains about $1\frac{1}{2}$ acres. (4840 sq. yds. = 1 acre.)

7. *ABC* is an equilateral $\triangle$ and from any point *D* in *AB* lines *DK* and *DL* are drawn parallel to *BC* and *AC* respectively. Find the ratio of the perimeter of the parallelogram *DLCK* to the perimeter of the $\triangle$ *ABC*. You may determine this if you choose by measurement, but a general proof should also be given.

8. Find what $2x^2-3x+4$ becomes when $x+a$ is substituted for x, and show that if $2x^2-3x+4$ is subtracted from the result the remainder thus obtained is divisible by $4x-3$ if a is so small that its square may be neglected.

9. The square *ABCD*, measuring 3 cm. in the side rolls without sliding along the line *XY*, turning about *B*, *C* and *D* in succession. Draw (full size) the path of the point *A* from its position on the line *XY* till it again lies on *XY*, and calculate the length of this path.

10. A bed of coal 14 feet thick is inclined at 23° to the surface. Calculate the number of tons of coal that lie under an acre (4840 square yards) of surface. A ton of coal occupies 28 cubic feet. (The 14 feet is to be regarded as a measurement at right angles to the surface of the coal bed.)

II. PART I. SECOND PAPER.

[N.B.—*In Geometry the demonstrations and sequence need not be Euclid's.*]

1. The London County Council has contracted a debt of £53,761,638. What sum is required to pay 3% interest on this?

2. A donkey is tethered 5 yards from a long straight hedge, and it can graze at a distance of 13 yards from its tether. Find how much of the hedge it can nibble.

Construct a figure on the scale of $\frac{2}{10}$ inch to the yard, on squared paper, and estimate the area over which the donkey can graze by counting the squares included in it, or by calculation.

3. The mean temperature on the first day of each month showed on an average of 50 years the following values:

Jan. 1, 37°	May 1, 50°	Sept. 1, 59°
Feb. 1, 38°	June 1, 57°	Oct. 1, 54°
Mar. 1, 40°	July 1, 62°	Nov. 1, 46°
April 1, 45°	Aug. 1, 62°	Dec. 1, 41°

Represent these variations by means of a curve. (You may neglect the difference of length of different months.)

4. A man 5 feet 9 inches high standing 134·2 feet from the foot of a tower observes the elevation of the tower to be 30° 14′. Find the height of the tower.

5. Find the values of $81^{\frac{1}{2}}$, $81^{-\frac{1}{2}}$, $81^{1\cdot25}$, $81^{-1\cdot25}$. Justify the interpretations you put on $81^{\frac{1}{2}}$, and $81^{-\frac{1}{2}}$.

Is $10^{6\cdot3714}$ greater or less than a million? Give reasons.

6. A party of tourists set out for a station 3 miles distant and go at the rate of 3 miles an hour. After going half a mile one of them has to return to the starting point; at what rate must he now walk in order to reach the station at the same time as the others?

7. The volume of the frustum of a cone can be obtained from the formula $V=\frac{1}{3}h\ (a+b+\sqrt{ab})$, h denoting the height of the frustum, a and b the areas of the two ends. Calculate the volume, to the nearest cubic inch, when $h=4\cdot5$ in., $a=28\cdot5$ sq. in., $b=78\cdot6$ sq. in.

What does the formula become (i) when $a=b$, (ii) when $a=0$? Give the geometrical interpretation in each case.

8. *OAB* is a quadrant of a circle and *AC*, *BC* are tangents. Any line *KLMN* is drawn parallel to *OA* cutting *OB* at *K*, *OC* at *L*, the arc *AB* at *M*, and *AC* at *N*. Prove that *OKL* is isosceles and that $KL^2+KM^2=KN^2$.

9. In a triangle *ABC* the angles *B* and *C* are equal, and the tangent of each of these angles is $\frac{4}{3}$.

Determine either by use of tables or by construction of a figure the value of the third angle. If you work graphically draw a large figure.

10. A spherical glass vessel has a cylindrical neck 8 cm. long, 2 cm. diameter; the diameter of the spherical part is 8·5 cm. By measuring the amount of water it holds a boy makes out its volume to be 345 cub. cm. Find by calculation whether he is correct, taking the above as inside measurements and π as 3·1416. Take the spherical vessel and the cylindrical neck to be complete, neglecting the fact that they overlap.

III. PART I. THIRD PAPER.

[N.B.—*In Geometry the demonstrations and sequence need not be Euclid's.*]

1. A gramophone with 5 rolls of music costs £7, the same gramophone with 20 rolls of music costs £9. How much should be paid for the gramophone with 50 rolls of music?

2. Draw a rectangle 2 ins. long and 1 in. broad and construct a rhombus of equal area, having each of its sides 2 ins. long. Measure or calculate the angles of the rhombus.

3. Calculate to four significant figures the values of the following quantities:

$$\text{(i)}\ \frac{52{\cdot}45 \times 378{\cdot}4 \times {\cdot}02086}{87{\cdot}32 \times {\cdot}5844};$$

$$\text{(ii)}\ (1{\cdot}246)^{4{\cdot}196}.$$

4. A meteor is seen at a height of 450 ft. above the ground, one second later it is 328 ft. above the ground, and two seconds from the time of first observation it is 174 ft. above the ground.

Determine the values of a, b, c, if the formula at^2+bt+c represents for all values of t the height of the meteor t seconds after the instant of the first observation.

5. In a triangle ABC, $BC = 93$ yds., $\angle ABC = 59^\circ\ 19'$, $\angle ACB = 43^\circ\ 15'$. Calculate the length of AB.

Also find what error is made in the length of AB if the angle ACB is through a wrong measurement taken as $43^\circ\ 17'$.

6. The following rules for conversion of units are sometimes employed:—

Having given that 1 kilogram = 2·204 pounds, find the percentage error in each case

(i) to convert ounces to grams multiply by $\frac{85}{3}$.

(ii) to convert pounds to kilograms multiply by $\frac{5}{11}$.

7. A rectangular fold is to be made with hurdles 6 ft. long for 1000 sheep, allowing not less than 8 sq. ft. per sheep. Find the number of hurdles needed for the cases in which one side of the fold consists of 10, 11 . . . 20 hurdles. Make a diagram on the squared paper showing the total number of hurdles required for a given number in the side, and from the diagram find the smallest number of hurdles which will suffice.

8. Construct a $\triangle ABC$ given $\angle A = 90°$, $AB = 8{\cdot}4$ cm., $AC = 6{\cdot}3$ cm. Bisect $\angle BAC$ by a line meeting BC at O. Measure BO, OC, and thus determine the ratio BO/OC. Verify your result by calculation.

Prove the property on which you base your calculation.

9. Sand lies against a wall, covering a strip of ground 4 ft. wide. If the sand will just rest with its surface inclined at 30° to the horizon, how much sand may lie on this strip per foot length of wall? Give the quantity to the nearest tenth of a cubic foot.

10. XB is the projection of AB on MN, the angle AXB being a right angle. Find the length of XB when $AB = 5$ ins. and the angle ABX (α) is equal to 33°.

If AB and BC are the sides of a square, and XB and BY their projections on MN, how must the square be placed for XY to have (i) the least, (ii) the greatest possible length, consistently with the conditions that B is always to be on MN and the square is to be above MN and in the same plane with it?

IV. PART II. FIRST PAPER.

1. In a figure of a cylindrical oil drum (drawn on a scale of $\frac{1}{10}$) the radius is 1·8 in. and the length 3·15 in. How many gallons will the drum hold? A gallon is 277 cubic ins.

Assuming that each of your measurements may be wrong by $\frac{1}{80}$ of an in., find how much per cent. your result may be wrong.

2. The distances and second class railway fares from Paris to the under-mentioned stations are as follows, the distances being reckoned in miles and the fares in francs and centimes (1 franc = 100 centimes):—

	Distance in Miles.	Fare. fr.	c.
St. Denis,	4	0	55
Chantilly,	$25\frac{1}{2}$	3	10
Creil,	32	3	85
Longueil,	45	5	45
Compiegne,	53	6	35
Noyon,	67	8	15
Chauny,	77	9	35
Tergnier,	82	9	90
St. Quentin	96	11	65

From these data draw on squared paper a diagram showing the relation between the distance travelled from Paris and the railway fare. Examine whether the railway fare is proportional to the distance travelled, assuming distances to be given to the nearest half mile, and that 5 centimes is the smallest coin used for payment of railway fares.

[*N.B.*—A diagram is meaningless unless the scale of measurement is stated and the positions of the lines from which distances are measured are indicated.]

3. If a square post has to be cut from a cylindrical tree trunk, find roughly what percentage of wood is wasted.

4. What are the values of

$$A\frac{(x-b)(x-c)}{(a-b)(a-c)}+B\frac{(x-c)(x-a)}{(b-c)(b-a)}+C\frac{(x-a)(x-b)}{(c-a)(c-b)}$$

when x is equal to a, b, and c respectively?

Find a quadratic function of x which is equal to 1 when x is 0, equal to 0 when x is 1, and equal to 2 when x is 2.

5. An upright pole 10 ft. high casts a shadow 12 ft. 6 in. long at mid-day on a certain day. Another upright pole of the same height, 100 miles further north, casts a shadow 13 ft. 2 in. long at the same time. Deduce the earth's perimeter, supposing the earth a sphere.

6. O is the centre of a circle of radius $2r$, and AOB a diameter, and circles are described on OA and OB as diameters. Another circle is to be drawn to touch these three circles. Show that the equation

$$(r+x)^2=(2r-x)^2+r^2$$

is true, where x denotes the radius of the required circle. Determine x in terms of r, and carry out the construction when $AB=12$ cm.

7. Assuming expressions for $\sin(A+B)$ and $\cos(A+B)$, find an expression for $\tan(A+B)$ in terms of $\tan A$ and $\tan B$.

The sights of a gun are two feet apart, and the back sight is raised till it is 2 ins. above the front sight when the barrel of the gun is pointing horizontally. I raise the gun till the line of sights points directly towards the top of a tower 100 ft. high and 500 yds. distant. Find the tangent of the angle of elevation at which the barrel points, and hence calculate the angle.

8. A semi-circular arch is built of stones, whose shape is a trapezium with parallel sides 7 in. and 6 in., and the other two sides each 1 ft. What is the internal diameter of the arch, and how many stones are required to complete it?

9. Two conics L and M have the same focus S and directrix, the length of the perpendicular SX from the focus on the directrix being $1\frac{1}{2}$ ins. The eccentricity of L is $\frac{1}{2}$, and that of M is 2. Draw a diagram showing the general shapes of the curves.

10. Show that either tangent to a conic from a point on its directrix subtends a right-angle at the focus.

WOOLWICH ENTRANCE EXAMINATION.

V. PART II. SECOND PAPER.

[*Take the acceleration of a falling body as 32 ft. per second per second.*]

1. Find *graphically* the resultant of the two forces 5 lbs.-weight and 4 lbs.-weight acting at an angle of 126°. Measure and write down the angle the resultant makes with the direction of the larger force. To obtain full marks the resultant must be measured accurately to the first place of decimals and the angle to the nearest degree.

2. Two smooth pegs *A* and *B* support a cord which measures 10 ft. between the pegs, and carries a weight of 20 lbs. by means of a second cord knotted to it at a point 4 ft. from the end attached to the peg *A*. *B* is 6 ft. to the right of, and 1·1 ft. vertically above *A*. Draw the cord in position and find the tension in each of the two parts into which the knot divides it.

3. A bar of uniform section 12 ft. long, and weighing 20 lbs. per foot, rests on two supports 9 ft. apart (the ends projecting equally beyond the supports), and carries a load of 300 lbs. at a distance of 3 ft. 6 ins. from the left hand support. Calculate,

(i) the pressure on each support due to the load carried;

(ii) the pressure on each support due to the combined effect of the load carried and the weight of the bar.

4. *A* and *B* is a bent lever whose arms $2\frac{1}{2}$ ft. and 1 ft. 7 in. long are inclined to one another at an angle of 118°. A force of 56 lbs. weight is applied horizontally to the long arm of the lever; find what force would be required to balance it acting vertically at the end of the short arm. Find the resultant of the two forces graphically.

5. A body is projected upwards with a velocity of 200 ft. per second. Find the height to which it will rise and the time it takes to arrive at its highest position.

6. The thin plate shown in a figure is uniform in thickness and density. The figure is made up of 3 rectangles whose dimensions are $2'' \times 3''$, $2'' \times 1''$, and $2'' \times 1\frac{3}{4}''$, placed side by side in such a way that the line *AB* of length 6″ bisects each rectangle. Find its centre of gravity and mark its position carefully on the figure.

7. A mass of 1,000 pounds is moving in a straight line at a speed of 30 miles per hour. Its energy is reduced to 500 ft. lbs. in 12 seconds. Calculate the retardation and the force which acted to produce it.

8. A force of 50 poundals acts upon a mass of 700 pounds for 10 seconds. Calculate the momentum possessed by the body at the end of the interval and also its kinetic energy, being careful to write down the units in which the results are expressed.

9. A lever safety valve turns about the fulcrum *A*, the resultant thrust of the steam on the valve acts vertically upwards at the point *B*, and the valve is held down by a weight applied at *C*. If *C* is 20 ins. from *B*, and *B* is 3 ins. from *A*, calculate the weight which must be applied so that the valve may just blow, when the upward thrust on the valve is 700 lbs. wt.

10. An ordinary winch is used to lift water out of a well. It is found that a load of 112 lbs. hanging on the rope can just be balanced by a force of 30 lbs. acting at the handle in a direction tangential to its circular path. The length of the handle is 20 ins. and the diameter of the barrel of the winch is 8 ins. Calculate the efficiency of the winch.

VI. PART III. First Paper.

[*Take the acceleration of a falling body as 32 foot-second units.*]

1. Determine graphically, or by calculation, the co-ordinates of the points of trisection of the line joining the points (8, 3) and (1, 7).

2. Prove that the straight line

$$5x+12y-4=0$$

touches the circle

$$x^2+y^2-6x+4y+12=0.$$

Determine the coordinates of the point of contact.

3. Find the coordinates of the centre of an ellipse shown in a diagram.

4. State the more important properties of the foci of an ellipse.

Given the foci and one point on the curve, show how it may be described mechanically.

5. Obtain the equations of the asymptotes of the hyperbola

$$x^2-4y^2+x+4y=0.$$

Draw the curve to scale on ruled paper, taking two inches as unit.

6. Prove that whatever m may be, the straight line

$$y=mx+\frac{1}{m}$$

touches the parabola

$$y^2=4x.$$

7. A block of stone weighing 400 lbs. lies on a floor, and is to be moved by a rope attached to it. The angle of friction between the stone and the floor is 35°. Determine graphically the force required to move the stone when the rope is pulled (1) horizontally, (2) at an inclination of 20° to the horizontal, (3) at an inclination of 40° to the horizontal.

By general reasoning, when the angle of friction is $\alpha°$, determine the direction in which the rope must be pulled to make the force required to move the stone as small as possible.

8. A gun weighing 4 tons, fires a shot of 160 lbs. If the powder generates 10^7 foot-pounds of mechanical energy, determine the muzzle velocity of the shot to the nearest 10 feet per second. Assume the gun horizontal and free to recoil.

9. The positions of the (vertical) legs of a three-legged table are indicated by the angular points of the triangle ABC. Neglecting the weight of the table, and supposing a weight of 100 lbs. to be placed on it at D, determine the thrust in each leg.

In the diagram $AB = 5$ in., $BC = 3\frac{3}{4}$ in., $CA = 3$ in., and the distances of D from these sides respectively were 1·03, ·88, ·91 inch.

VII. PART III. SECOND PAPER.

[*Take the acceleration of a falling body as 32 foot-second units.*]

1. Draw on the squared paper the line bisecting at right angles the line which joins the points (5, 1) and (2, 4), unit 1 inch.

By inspection of your figure write down the equation of the line so found, and verify your result by working out the equation.

2. Find the equation of the locus of the centres of circles passing through the point (1, 2), and touching the line $y = 0$.

Indicate, by drawing a rough diagram, the position and the nature of the locus.

3. With a unit of 1 inch draw the curve given by $x = \sec\theta$, $y = \tan\theta$, when θ varies.

4. With a unit of 0·2 inch plot the four points (16, 1), (16, 3), (14, 7), (2, 3). Show by calculation that they are concyclic, and draw the circle on your diagram.

5. In a Diagram mark 2 points A, A' on the axis of x distant $1\frac{1}{2}$ inches from the origin, and 2 points B, B' on the axis of y distant $\frac{3}{4}$ inch. Draw a line PP' making intercepts $3\frac{3}{16}$ and $\frac{17}{20}$ inches on the axes. If PP' is a tangent to the ellipse whose axes are the lines AA' and BB', find the foci and the point of contact of the tangent. State the steps of your construction.

6. A train of mass 300 tons moves up an incline of 1 in 120, at a uniform speed of 40 miles per hour. The frictional resistances amount to 5 poundsweight per ton. Find the horse-power developed by the engine.

7. A and B are two fixed pulleys, C a movable one. A string ACB passes over A and B and supports C. At the free ends of the string weights of 5 pounds and 3 pounds respectively are attached. The pulley C with weight attached weighs 9 pounds. Find the accelerations of the weights.

8. A pair of rafters, inclined to the horizon at an angle of 37° rest on walls and support a roof. The roof is equivalent to a load of 200 pounds-weight at the middle of each rafter. The rafters may be considered freely hinged to one another and to the walls, and their own weights may be neglected. Find graphically the thrust on each wall.

If the lower ends of the rafters are to be joined by a rod to relieve the walls of horizontal pressure, what will the pull on the rod be?

9. A particle moves in a straight line with uniform acceleration of 0·5 foot-second units, the initial velocity being 0·7 feet per second. Construct on the squared paper a diagram, in which the ordinate represents the space passed over in a given time, 2 inches on the paper representing a foot, and also representing a second.

Show that the tangent of the inclination to the time axis of a line through the points on the diagram corresponding to $\frac{9}{10}$ and $1\frac{1}{10}$ seconds is equal numerically to the velocity at the end of 1 second. Give reasons for this equality.

ANSWERS.

JUNE, 1894.

II. Arithmetic.

1. $22\frac{5}{7}$. 2. ·0016. 3. £74. 11*s*. $10\frac{1}{2}$*d*.

4. £1. 13*s*. $10\frac{1}{2}$*d*. 5. $\frac{7}{40}$. 6. £30. 12*s*. 3*d*.

7. 660. 8. $2\frac{1}{2}$ inches. 9. 12.

11. *A* had originally £1380, *B* £3220. 12. 6 days.

13. 15. 14. £5.

III. Algebra.

1. $x^3-3x^2+11x-8$.

2. $(x-2y)(x^2+2xy+4y^2)$, $(x-8y)(x-16y)$, $(x^2+xy+y^2)(x^2-xy+y^2)$.

3. x^2-7x+2. 4. $5(a^2+b^2+c^2-bc-ac-ab)$.

5. (i.) n^2+3n+1; (ii.) $\sqrt{13}+\sqrt{11}$. 6. (i.) $\frac{a+c}{2}$; (ii.) 4, −7.

7. $x^2+3px+2p^2+q=0$. 8. 16 and 40 years.

9. 12, $\frac{1}{12}$. 12. 2·5854607, ·1602526, 5·82.

IV. Trigonometry.

1. $343° \, 38' \, 10\frac{10}{11}''$. 2. $-\frac{2}{3\sqrt{7}}$.

3. 49·177 ft. 4. $-\frac{1}{\sqrt{2}}$, $-\frac{\sqrt{3}}{2}$, -1, $\frac{2}{\sqrt{3}}$.

8. $21° \, 47' \, 12\frac{1}{2}''$, $38° \, 12' \, 47\frac{1}{2}''$, $120°$.

9. $$\left\{\frac{x^2\sin^2\alpha}{\sin^2(\beta-\alpha)}-\frac{2xy\sin\alpha\cos\beta\sin\gamma}{\sin(\beta-\alpha)\sin(\gamma-\beta)}+\frac{y^2\sin^2\gamma}{\sin^2(\gamma-\beta)}\right\}^{\frac{1}{2}}.$$

10. 60°, 44° 49′ 20″. 11. 114582 sq. ft.

12. 4·0825 inches.

WOOLWICH ENTRANCE EXAMINATION.

V. Statics and Dynamics.

1. (i.) Equilibrium impossible; (ii.) Parallel to sides of a right-angled triangle.

2. $\sqrt{\{1440+304\sqrt{2}+180\sqrt{3}-144\sqrt{6}\}}$, $\tan^{-1}\dfrac{29+8\sqrt{2}}{10+9\sqrt{3}-8\sqrt{2}}$ with OA.

3. (i.) $216\frac{2}{3}$ lbs. (ii.) 200 lbs.

5. 5 lbs.-weight at distance 16″ from greater force.

6. $\frac{25}{36}$ inch from centre. 8. Ratio is 11 : 10.

9. At height 3520 ft., 11 secs. after the first was fired.

10. 3 lbs.-weight, or 96 poundals.

11. Range 7500, velocity $20\sqrt{241}$ in direction $\tan^{-1}\frac{4}{15}$ to horizontal.

VI. Pure Mathematics.

2. ·000762588, ·000000566236. 3. $\frac{25}{27}$ secs., $1\frac{2}{23}$ secs.

9. $\dfrac{x}{p}+\dfrac{y}{q}=2.$

10. If the equation to the circle be $x^2+y^2+Ax+By+C=0$, $(A+B)^2>8C$.

11. $x^2+y^2-5ax=0$; $\left(a,\ 2a\right)$, $\left(\dfrac{9a}{2},\ -\dfrac{3a}{2}\right)$, and $\left(a,\ -2a\right)$, $\left(\dfrac{9a}{2},\ \dfrac{3a}{2}\right)$.

13. $x^2+y^2+2xy\dfrac{a^2-b^2}{a^2+b^2}=a^2-b^2.$

VII. Mechanics.

2. $n\pi$. 3. $\dfrac{3b}{\sqrt{2}}$. 5. $\frac{3}{4}\,l$. from end of the bar.

8. 20,000 ft., 25·5 ft. 9. $\frac{2}{3}W$.

NOVEMBER, 1894.

II. Arithmetic.

1. $\frac{1}{8}$. **2.** 440·82. **3.** $4\frac{1}{2}$ per cent.

4. 4 cwt. 1 qr. 14 lbs. **5.** $\frac{1}{60}$. **6.** £200. 10*s*. $2\frac{3}{4}d$.

7. 47. **9.** £80. **10.** 3 p.m.

12. Latter. **13.** £194. 5*s*., £224. 14*s*.

14. $37\frac{1}{2}$ miles.

III. Algebra.

1. $x^4+ax^3-a^3x-a^4$. **2.** $-\frac{(x+y-z)^2}{2yz}$.

4. $-(a+b+c)$. **5.** $x=-3,\ y=1\frac{1}{2},\ z=4$.

6. $-\frac{1}{2},\ \pm 3$. **7.** $-1,\ -\frac{1}{2}$.

8. Series is $3+2\frac{1}{2}+2+1\frac{1}{2}+\ \dots$.

9. $1+\frac{1}{3}x+\frac{2}{9}x^2+\frac{14}{81}x^3+\ \dots\ +\frac{1\cdot 4\cdot 7\dots(3n-5)}{3^{n-1}\lfloor n-1}x^{n-1}+\ \dots$.

10. 45 and 36 miles an hour. **11.** 3·2101493, 2·30258.

IV. Trigonometry.

1. $57\frac{3}{11}^\circ$. **3.** 15°. **5.** 20°.

8. $B=54^\circ\ 56'\ 48''$, $C=83^\circ\ 3'\ 12''$, $c=209{\cdot}174$; or $B=125^\circ\ 3'\ 12''$, $C=12^\circ\ 56'\ 48''$, $c=47{\cdot}2108$.

9. $109^\circ\ 6'$. **10.** $3\frac{1}{7}$ sq. feet. **11.** 12·143 sq. inches.

V. Statics and Dynamics.

1. $2AC$. **3.** 100 lbs., 40 lbs.; 112 lbs., 28 lbs.

5. $\frac{W}{2},\ \frac{W}{\sqrt{3}}$. **6.** 1

7. 1024 ft., 16 seconds; 4 and 12 seconds.

9. 45°, 160 f.s. **10.** 2 lbs.-weight, or 64 poundals.

WOOLWICH ENTRANCE EXAMINATION.

VI. Pure Mathematics.

4. 103923, percentage ·7725. 5. 4.

6. 82° 25′ 22″ and 57° 34′ 38″.

7. $\frac{4}{p}\sin\frac{\phi''-\phi'}{2}\sin\frac{\phi''-\phi}{2}\sin\frac{\phi'-\phi}{2}$, ·000227633. 10. $y-k=\sqrt{3}(x-h)$.

11. $2\sin^{-1}\frac{\rho}{\sqrt{a^2+\beta^2}}$. 14. $(\sqrt{2}-1)x^2-(\sqrt{2}+1)y^2=\frac{25}{2\sqrt{2}}$.

VII. Mechanics.

1. 7·89. 2. $\frac{W\sqrt{3}}{2}$ making angle 60° with horizon.

3. $\sqrt{13}:2$. 5. 14 lbs. 8. 100 ft.

9. $3\frac{1}{2}$ seconds. 10. $50\sqrt{2}$ f.s., $\tan^{-1}\frac{1}{3}$.

ANSWERS.

JUNE, 1895.

II. Arithmetic.

1. $30\frac{13}{18}$. **2.** ·22. **3.** $7\frac{1}{2}$ years.

4. ·7875. **5.** £74. 1*s*. $5\frac{1}{2}$*d*. **6.** £555. 10*s*. 6*d*.

7. 325. **8.** $6\frac{1}{4}$ tons. **9.** 30.

11. *A*, £540; *B*, £675; *C*, £1200. **12.** 3 per cent.

13. 55, 25. **14.** £23. 14*s*.

III. Algebra.

1. 0. **2.** (i.) $(x-1)^2(x+1)(x-2)^2(x^2-2)(x^2+2)^3$; (ii.) $x^8(x^8-y^8)$.

4. (i.) 8, 9; (ii.) x, $7\frac{2}{11}$; y, 3; (iii.) $x=\pm\frac{1}{2}, \pm\frac{4}{3\sqrt{3}}$; $y=\pm\frac{1}{3}, \mp\frac{7}{6\sqrt{3}}$.

5. *A*, 67; *B*, 28. **9.** $b^y=c^x$, 1·77245.

IV. Trigonometry.

1. ·8108936, 1·1934918. **2.** (i.) $\frac{15}{17}$; (ii.) $\frac{5}{\sqrt{34}}$.

3. $\frac{y}{x}$ and $\frac{x}{y}$. **9.** 53° 7′ $48\frac{1}{2}$″. **10.** 15 ft.

V. Statics and Dynamics.

1. 0. **4.** $\frac{1}{6}$ weight of lever. **5.** 10,000 lbs.

6. $\frac{1}{120}$ f.s., $\frac{13}{240}$ f.s. **7.** (i.) $\frac{3}{2}$ f.s.s.; (ii.) 2 f.s.s.

8. (i.) 2 f.s.s.; (ii.) $16\sqrt{3}$ f.s.; (iii.) $8\sqrt{3}$ sec. **9.** $\frac{11}{6300}$. **10.** 30°.

VI. PURE MATHEMATICS.

1. (i.) a^2, a; (ii.) $x = \pm\frac{1}{2}(\sqrt{3b^2-2a^2} \pm \sqrt{2a^2-b^2})$; $y = \frac{1}{2}(\sqrt{3b^2-2a^2} \mp \sqrt{2a^2-b^2})$.

5. 173·089 ft. **8.** 9, 6; −6, −4.

9. $11x + 3y = 30$, $3x - 11y = 20$.

10. Radius 1, points of contact, 1, 1 and 2, 2.

12. The co-ordinates of the point are $c \tan(\theta+\phi)$, $-c \tan(\theta+\phi)$.

14. $GK = \frac{1}{2}$ latus rectum.

VII. MECHANICS.

1. 7 lbs. **2.** $33\frac{5}{24}$, $33\frac{19}{24}$ lbs.

3. Distance from centre $= \frac{1}{18}$ side of square. **4.** $\frac{28}{171}$.

5. 6 ft. **6.** 12 : 5. **7.** $\frac{7}{48}$ mile, 35·35 miles per hour.

9. 60 f.s., $\tan^{-1}\frac{3}{4}$. **10.** $W\left(\cos\alpha + \frac{v^2}{gL}\right)$, $W\left(\sin\alpha - \frac{v^2}{gL}\cot\alpha\right)$.

NOVEMBER, 1895.

II. Arithmetic.

1. $\frac{3}{8}$. 2. $156\dot{2}{\cdot}5$ 3. £531. 13*s.* 4*d.*
4. 31 lbs. 4 oz. 16 dwts. 16 grs. 5. ·941$\dot{6}$. 6. £1092. 15*s.* 8*d.*
7. 1155. 8. £7. 15*s.* 9. £25. 13*s.* 4*d.*
11. $1\frac{1}{2}$ inches. 12. £17. 10*s.*, £70, £157. 10*s.*
13. $6\frac{3}{7}$ secs., 204 yds. 14. 7 per cent.

III. Algebra.

1. $x^3 - 2x^2 + 4x + 7$.
2. (i.) $(x+7)(x+9)$; (ii.) $(y-a)(y+7a)(y-6a)$;
   (iii.) $x(x+1)(x-1)(x+2)(x-2)(x+3)(x-3)$.
3. $x^2 - 5x + 1$. 4. $(a+b-c)(a-b+c)(a-b-c)$.
5. (i.) 6, 5; (ii.) $\frac{4}{5}$, $-\frac{3}{5}$; (iii.) $x = \pm 2, \pm 1$; $y = \pm 1, \mp 2$. 6. £4000.
7. 51 miles an hour, $93\frac{1}{2}$ miles. 8. $2\frac{179}{243}$, 45 terms.
9. $4(p^2 - q)$. 10. 136080.
11. 1, 1, $\frac{3}{2}$, $\frac{5}{2}$, $\frac{35}{8}$, $\frac{63}{8}$; greatest term $= 9289\frac{91}{128}$. 12. ·418518.

IV. Trigonometry.

1. (i.) 144°; (ii.) $\frac{4\pi}{5}$. 2. $\frac{3}{4}\sin^2 2A$. 3. 10°.
4. $n\pi + \frac{\pi}{3}$, $n\pi + \frac{3\pi}{4}$. 9. 205·4.
11. Area ·0403125 sq. inches; length, 6·1416 inches.

V. Statics and Dynamics.

1. 135°. 2. Length, 20 inches; weight, 12 lbs.
3. Force, 2 lbs.; hinge action, $2\sqrt{5}$ lbs., making angle $\tan^{-1}\frac{1}{2}$ with AB.
4. $\frac{7\sqrt{3}}{10}a$ from vertex of lightest triangle.
5. Second system, 4 sheaves in upper, 3 in lower block, with the string fastened to lower block. 35 ft.
6. 10·6 miles an hour. 8. 312 ft.; 64 f.s., 144 f.s.
9. $\frac{5\sqrt{2}}{6}$ seconds. 10. $\frac{1}{6}$ range.

WOOLWICH ENTRANCE EXAMINATION.

VI. Pure Mathematics.

1. £5436. 2*s.* $5\frac{1}{2}$*d.*

2. *A* arrives 1 min. 24 secs. before *B*; $12\frac{1}{2}$ miles an hour.

3. ·47829, ·96059.

4. $-\tan\frac{\alpha+\beta}{2}$, $\dfrac{\cos\frac{\alpha-\beta}{2}}{\cos\frac{\alpha+\beta}{2}}$.

5. $a=\frac{2}{3}\sqrt{2q^2+2r^2-p^2}$,
   $\cos A=(5p^2-q^2-r^2)\div 2\sqrt{(2p^2+2q^2-r^2)(2p^2+2r^2-q^2)}$,
   and similar expressions. The error in the numerator of $\cos A$ is $18x$.

8. (3, 2), (−3, −2), (2, 3), (−2, −3).

9. $3x=4y$.

10. $x^2+y^2-18x-8y=3$.

VII. Mechanics.

1. $P\sqrt{2}$ acting at 45° to horizontal.

2. 220 lbs. It will rest under the action of the friction alone.

3. *P* has any value from 6 to infinity.

4. $\tan^{-1}\frac{3}{10}$, $7\frac{1}{2}$ ft.

5. Former, $9\frac{1}{8}$ ft.

6. $6\frac{1}{4}$ seconds.

7. (i.) 8 oz.; (ii.) $8\frac{1}{2}$ oz.

8. 5 and 15 f.s., 900 oz.-wt.

9. 2 oz.-wt. towards centre.

10. $23\frac{1}{3}$ secs., $5288\frac{8}{9}$ ft. up the plane.

ANSWERS.

JUNE, 1896.

II. Arithmetic.

1. $13\frac{1}{14}$.
2. 100·37.
3. $7\frac{1}{2}$ years.
4. 2 fur. 18 poles 4·4 yds.
5. $\frac{7}{40}$.
6. £1657. 16*s*. $6\frac{1}{4}$*d*.
7. 6105.
8. £4. 17*s*. 6*d*.
9. £5.
10. 3 qrs. 23 lbs.
11. £176, £320, £432.
12. 8 ft.
13. 10 cwt. zinc, 4 cwt. copper.
14. 15 miles an hour.

III. Algebra.

1. $367a - 114b + 690c$, 1082.
2. $x^6 - y^6$.
3. $-(b-c)(c-a)(a-b)(a+b+c)$, $(x-1)(x+1)(x+2)(x+3)$.
4. H.C.F $x^2 - 10x + 21$, L.C.M. $(x^2 - 10x + 21)(x-1)(x-2)(x-4)(x-5)$.
5. 1300.
6. (i.) 1, 5; (ii.) 1, 1, 3, $\frac{1}{3}$; (iii.) $9x^2 - 27x + 20 = 0$.
7. 3.
9. $66^3 = 287496$.
10. 4368.
11. 14, 6770 ....
12. (i.) $\frac{3}{4}$; (ii.) Conservative 12944, Liberal 11120; (iii.) 21056 each.

IV. Trigonometry.

1. 3.
3. $2\cos A$, $-\cos A \pm \sqrt{3}\sin A$.
8. $90°$ and $15° 15' 36''$.
9. 1026·337 ft.
11. (i.) $3\frac{19}{375}$ cub. inches; (ii.) 15·65 acres.

V. Statics and Dynamics.

1. 11 lbs. 5. $\frac{1}{3}W$. $\sqrt{\frac{10}{3}}W$.

6. Divides it in the ratio 2 : 5. 7. 96. 8. N.E. or N.W.

9. 64 ft.; time of ascent or descent 2 secs. $\frac{1}{2}$ and $3\frac{1}{2}$ secs.

10. 2 lbs.-weight. 11. $v^2 : gr$.

VI. Pure Mathematics.

1. 2 ft. 11·2 inches; 2500 : 2401. 3. 9000.

6. $x^2+x-1=0$. 8. $l\cos(\alpha-\beta)=p\pm a$.

9. $\frac{ar'+a'r}{r'-r}$, 0, and $\frac{ar'-a'r}{r'+r}$, 0. $2(a'+a)x=r'^2-r^2-a'^2+a^2$.

10. Write $-\sin 2\theta$ for $\sin 2\theta$. 11. $\frac{\sqrt{3}}{4\sqrt{2}}$, $-\frac{\sqrt{3}}{2\sqrt{2}}$.

13. A parabola with same axis and vertex at the focus of the given parabola.

14. A circle centre on the minor axis distant $CS\cot\alpha$ from the centre and radius $a\operatorname{cosec}\alpha$, where α is the given angle.

VII. Mechanics.

2. 3·38, 2·19 oz.-weight. 3. Tension $\frac{7}{40}W$, Pressures $\frac{3}{5}W$, $\frac{47}{40}W$.

4. ·634 AB. 5. $\mu^3-\mu^2-3\mu-1 : \mu^3+\mu^2-3\mu+1$.

6. It begins $220\frac{1}{2}$ ft. from the start.

7. Tension $6\frac{6}{13}$, Pressures $2\frac{10}{13}$, $2\frac{2}{13}$ oz.-weight.

8. $72(x-a)+23y=0$ and $89y=72(x+a)$.

NOVEMBER, 1896.

II. Arithmetic.

1. $13\frac{1}{8}$. 2. 28920. 3. 3570, $\frac{1}{8}$.

4. 10*s*. 7*d*. 5. £24. 9*s*. $6\frac{5}{18}$*d*.

6. 96 yds., 29 yds. 1 ft. $\frac{33}{56}$. 7. $3\frac{1}{3}$%.

8. $33\frac{1}{3}$ ft. 9. *A* £7. 10*s*.; *B* £6. 15*s*.; *C* £10. 10*s*.

11. 2464 cubic ft. 12. 7 : 5. 13. £41. 13*s*. 4*d*.

14. 18 and 10 miles a day.

III. Algebra.

1. $(x^2+y)(x^4-x^2y+y^2)$. $\frac{1}{16}$.

2. $(1+x)(1+x^2)(1-x-x^2-x^3+x^4)$. $3x^2-5x+7$.

3. $\frac{x^2-xy+y^2}{x^3-x^2y+xy^2-y^3}$.

4. (i.) $6\frac{1}{4}$, $2\frac{7}{8}$; (ii.) $x=\pm\frac{1}{2}$, $\pm\frac{4}{\sqrt{47}}$; $y=\mp\frac{1}{3}$, $\pm\frac{3}{\sqrt{47}}$.

5. 105 lbs. 6. 8.

8. 520. 10. 23·0258.

IV. Trigonometry.

2. (i.) 60°, 240°; (ii.) 270°; (iii.) 0°, 180°.

3. $\sin 105° = \frac{\sqrt{3}+1}{2\sqrt{2}}$, $\cos 105 = \frac{1-\sqrt{3}}{2\sqrt{2}}$, etc.

4. $\frac{3}{4}$, $\frac{9}{16}$, $\frac{1}{8}$; 41° 25′. 5. *B* 107° 57′ 49″, *C* 15° 2′ 11″.

10. 4 . radius.

V. Statics and Dynamics.

2. *AB*. 3. (i.) 10 lbs.; (ii.) 15 lbs.

5. *W*. 6. Power is (i.) $\frac{W}{2}$, (ii.) $\frac{W}{\sqrt{2}}$.

7. 2 : 1. 8. (i.) *W*; (ii.) $\frac{4}{5}W$. 10. 12 ft., 4 ft.

11. Its distances from *AB*, *AC* are 16, 32 ft., and it is moving in a direction equally inclined to *AB*, *AC*.

VI. Pure Mathematics.

2. $\frac{1}{50}(n+5)(n+10)$, $-\frac{3}{50}(n+4)(n+9)$, $\frac{3}{50}(n+3)(n+8)$, $-(n+2)(n+7)$, 0 according as n is of the form $5r$, $5r+1$, $5r+2$, $5r+3$, $5r+4$.

7. 2743·86 ft.

8. Taking centre of pentagon as origin, and a line joining it to one angular point as the axis of x, the equations are

$$x \cos 36^\circ \pm y \sin 36^\circ = \frac{a}{2} \cot 36^\circ, \quad x \sin 18^\circ \pm y \cos 18^\circ + \frac{a}{2} \cot 36^\circ = 0,$$

$$x + \frac{a}{2} \cot 36^\circ = 0, \quad 4(x^2+y^2) = a^2 \cot^2 36^\circ.$$

9. $\frac{\pi}{2} - a$, $\frac{n^2 \cot a}{l^2 - 2lm \operatorname{cosec} a + m^2}$.

11. A concentric ellipse semi-axes $\frac{ab}{\sqrt{a^2+b^2}}$, b.

VII. Mechanics.

1. $2k \,.\, AD$.

2. 38·78 and 17·22 lbs.

4. $7\frac{91}{99}$ lbs.

6. 40 miles an hour, $\frac{1}{24}$ f.s.s.

7. 4 f.s.s., 9 inches.

8. $\frac{4mn^2\pi^2 r}{g}$, $76\frac{4}{11}$.

9. After 1″ (32, 64); 2″ (64, 96); 3″ (96, 96); 4″ (128, 64); 5″ (160, 0); focus (80, 84).

10. $\frac{13}{64}U$, $\frac{15}{64}U$, $\frac{9}{16}U$.

ANSWERS.

JUNE, 1897.

II. Arithmetic.

1. £670. 8s. 0$\frac{3}{4}$d. 2. ·0051. 3. 437. 4. $\frac{5}{18}$.
5. $\frac{4}{15}$. 6. 14·552. 7. £688. 10s. 8d. 9. 70 days.
10. £20. 11. 25s. 12. 7 : 5. 13. 49$\frac{93}{343}$.
14. £600, £594, £609, £621, £596, £646.

III. Algebra.

1. $-(yz+xz+xy)$, $(1-b)(1-a)(1-ab)(1+a+ab)$.
2. $6x+13$, 9709, 23287, 73. 3. 1.
4. (i.) $-\frac{4a}{3}$; (ii.) $x=y=z=-\frac{1}{3}$; $x=\frac{1}{3}$, $y=z=-\frac{2}{3}$; $y=\frac{1}{3}$, $x=z=-\frac{2}{3}$; and $z=\frac{1}{3}$, $x=y=-\frac{2}{3}$.
5. In the larger, wine : water $=p^2-32p+1280 : 32p-p^2$, and in smaller, $32-p^2 : p^2-32p+320$. $p=16$.
6. 1296 and 2025, 2025 and 3025, 3025 and 4356, etc., $13^3=91^2-78^2$.
8. (i.) $2^{30}-31931$; (ii.) $2^{29}-27841$. 10. 6·943.

IV. Trigonometry.

1. 3·1416. 4. $n\pi+(-1)^n\frac{\pi}{6}$. 6. 500. 7. 567·138 yds.
8. 54·53516 sq. in. 9. 0, −1.
10. (i.) 1232 cub. ft.; (ii.) $\sqrt{6} : \sqrt{\pi}$.

V. Statics and Dynamics.

1. 11 lbs. 15 oz. and 11 lbs. at the ends; 11 lbs. 5 oz. and 11 lbs. 10 oz. at the points of trisection.
2. 6 oz. 3. 85 lbs. 5. 4 in. 7. 45 miles an hour, or 66 f.s.
8. 3 secs., 96 f.s., 72 ft.-lbs. 9. 4 secs., 64 f.s.
10. 2 secs. 11. 16 ft., $64\sqrt{3}$ ft.

VI. Pure Mathematics.

2. O is the orthocentre of the triangle.

3. (i.) when n is a positive integer; (ii.) when x lies between 1 and -1, when $x=1$ and $n>-1$, and when $x=-1$ and $n>0$.

4. $\frac{1}{4}(b \pm \sqrt{32a^2-b^2})$, b must not be $>4\sqrt{2}\,.\,a$.

6. $\frac{1}{2}(\operatorname{cosec} x - \operatorname{cosec} 3^n\,.\,x)$. **7.** 1889·63 ft.

8. $cx+(2b-c)y=bc$, $(b-c)x-(b+c)y=0$, $bx+(b-2c)y=bc$,
$(c-2b)x+cy+b^2=0$, $(b+c)x+(b-c)y=bc$, $(2c-b)x+by=c^2$,
where $b=AB$, $c=AC$.

9. A straight line.

10. $\dfrac{4a^2b^2}{(a^2+b^2)^{\frac{3}{2}}}$, where a, b are the lengths of the tangents.

11. $\dfrac{b^2}{\sqrt{(a^2\cos^2\phi+b^2\sin^2\phi)}}$.

VII. Mechanics.

1. (i.) $\dfrac{4r^2+h^2}{4(r^2+h^2)}\,.\,W$; (ii.) $\frac{1}{4}W$.

2. $Q+\frac{1}{2}(\sqrt{5}+1)(P-R)$, $P+\frac{1}{2}(\sqrt{5}+1)(Q-R)$.

3. $\dfrac{b-a}{6(b+a)}\,.\,h$. **4.** 13·66 lbs. **6.** 37½ ft. per hour.

8. $\tan^{-1}\frac{3}{4}$, $\tan^{-1}\frac{63}{16}$.

NOVEMBER, 1897.

II. ARITHMETIC.

1. $\frac{44}{135}$. **2.** ·000008. **3.** £17. 10*s.*, £9. 14*s.* $5\frac{1}{3}$*d.* **4.** $\frac{16}{165}$.

5. £46. 17*s.* 6*d.* **6.** $3\frac{1}{3}$. **7.** 162104 fr. $16\frac{3}{8}$ c.

9. £197. 17*s.* 4*d.* **10.** 3 ft. $2\frac{3}{8}$ in. **11.** 9·049 in.

12. *A*, 84*s.*; *B*, 72*s.*; *C*, 70*s.* **13.** 15 : 14.

14. 88 yds.; 22 miles an hour.

III. ALGEBRA.

1. $x = a^3 - 2a$, $y = 3a^2 - 1$, product $a^6 - 13a^4 + 10a^2 - 1$.

2. (i.) $a(x^2 + mx + m^2) + b(x + m) + c$; (ii.) $3xyz$;
(iii.) $3abc(b - c)(c - a)(a - b)$.

3. (i.) $(x^2 + 1{\cdot}414x + 1)(x^2 - 1{\cdot}414x + 1)$, (ii.) $(2x + {\cdot}628)(x + 3{\cdot}186)$.

4. (i.) $a^{-\frac{1}{3}}b^{-\frac{1}{2}}$; (ii.) 4.

5. (i.) $3\frac{1}{4}$; (ii.) $\pm a(la^2 + mb^2 + nc^2)^{-\frac{1}{2}}$, $\pm b(la^2 + mb^2 + nc^2)^{-\frac{1}{2}}$,
$\pm c(la^2 + mb^2 + nc^2)^{-\frac{1}{2}}$.

8. 76. **9.** $\frac{\lfloor 20}{32\lfloor 6}$, $\frac{13\lfloor 14}{32}$.

IV. TRIGONOMETRY.

1. $4\sqrt{5}$, $2\sqrt{5}$, $8\sqrt{5}$. **2.** (i.) $-$, $-$; (ii.) $+$, $+$.

3. $\frac{\sqrt{2} - 1}{\sqrt{6}}$, $-\frac{\sqrt{2} + 1}{\sqrt{6}}$. **4.** $(2n + 1)\pi$, $2n\pi \pm \frac{\pi}{3}$.

6. 3·90505 ft. **7.** 23·094 ft. **9.** $\frac{\pi}{4}$.

10. $\sqrt{1 + \sin\frac{a}{2}} : \sqrt{2} - \sqrt{1 + \sin\frac{a}{2}}$.

V. STATICS AND DYNAMICS.

2. $\frac{1}{3}W$, $\frac{2}{3}W$.

3. The stresses at *B* and *C* act in line *BC*, at *A* in line joining *A* to intersection of *BC* and *PQ*.

5. $W + 7w = 8P$. **6.** 352, 704 yards. **7.** 2 f.s.s.

8. 144 ft., 3 secs., 2 and 4 secs. **10.** 4 f.s.

WOOLWICH ENTRANCE EXAMINATION.

VI. Pure Mathematics.

3. £30. **4.** $496\frac{18}{25}$ miles, $\frac{17}{18}$ knot.

5. 2·30258, ·43429. **7.** 2·9 knots.

8. A circle, with the perpendicular from the focus on the tangent as diameter.

9. $(c\cos A - b)x + c\sin A \,.\, y = 0$, A being taken as origin and AB as axis of x.

10. A circle.

11. Taking as axes AB and the perpendicular from P on AB, the centre is $(c, 0)$, and the asymptotes are $x = c$, $y = 0$.

12. (i.) 2; (ii.) 3.

VII. Mechanics.

1. 9·6 lbs. making 37° with AB.

3. (i.) $\dfrac{2ab(4a^2 - b^2)}{(4a^2+b^2)^2} \,.\, W$; (ii.) $\dfrac{16a^4 + b^4}{(4a^2+b^2)^2} \,.\, W$; (iii.) $\dfrac{2ab}{4a^2+b^2} \,.\, W$.

6. 330 ft.-tons. **7.** 7 : 18. **9.** $\tan^{-1}\dfrac{k_2 + ek_1}{h_2 + eh_1}$ with AB.

10. $\dfrac{4}{\pi\sqrt[4]{5}} = {\cdot}85$; $\frac{3}{8}\sqrt{5}$, $\frac{9}{8}\sqrt{5}$ lbs.-wt.

ANSWERS.

JUNE, 1898.

II. Arithmetic.

1. $\frac{31}{48}$. 2. 42 sq. yds. 3. 3388 sq. yds. 4. $\cdot 0852\dot{2}\dot{7}$, $\cdot 0075$.

5. $\cdot 00161$. 6. £28. 1*s*. 7. 16380, 455 yds. 8. $\cdot 1\dot{3}$.

9. 6 cwt. 3 qrs. 41 lbs. 10. *A*, £680; *B*, £510; *C*, £340.

11. 168. 12. 53, 59, 61, 67, 71, 73, 79, 83, 89, 97.

13. 7 : 3. 14. 37.

III. Algebra.

2. $x-3$. 3. (i.) 0; (ii.) $a+b+c$; (iii.) $\frac{4a^2b^2}{a^6-b^6}$.

4. $a^{\frac{2}{3}}x^{\frac{7}{12}}(by)^{-\frac{3}{4}}$. 5. $(q-1)^2x^2-(q^2-1)dx+qd^2=0$.

6. (i.) $\frac{(b-c)^2}{8a(b+c)}$; (ii.) 1, 1; (iii.) $-5\frac{5}{12}$, 1.

8. $\frac{|\underline{10}}{16}=226800$; $4|\underline{7}=20160$; 323, of which 72 will be repetitions in amount, but paid by different coins.

9. 2^n, $\frac{|\underline{2n}}{(|\underline{n})^2}$.

10. When n is divisible by 4 the coefficient $=-\frac{3}{n}$, otherwise $\frac{1}{n}$.

IV. Trigonometry.

1. Long hand, 510° or $\frac{17\pi}{6}$; small hand, $42\frac{1}{2}°$ or $\frac{17}{72}\pi$.

2. 0°, 30°, 90°, 150°, 180°. 3. $2n\pi+\tan^{-1}\frac{b}{a}$.

6. 117″. 7. 2·72875. 9. 114·062 yds.

10. (i.) $\frac{l}{6}(ab+bc+cd)$; (ii.) $\frac{l^2}{6}(a+c+2b+2d)$, where a, b, c, d are the distances of the points where the plane cuts the horizontal edges from the corners of the cube, and l is the length of the edge of the cube.

V. Statics and Dynamics.

1. 60°. 2. $Q=2w$, $AD=\frac{1}{4}AB$. 3. 1 : 18. 4. 7·68 inches.

5. The weights are equally distant from the horizontal line through the centre. The greatest weight $=\frac{wK}{r}$.

7. Acceleration = 8.

8. Acceleration, 2 f.s.s. ; tension, $15\frac{15}{16}$ oz. ; pressures, $14\frac{1}{16}$, $13\frac{13}{16}$ oz.

9. After 4 seconds, it strikes the ground 512 ft. from the tower at an angle of 45°.

10. 12 f.s.

VI. Pure Mathematics.

4. $\frac{128}{27}$. 5. 28 years. 6. 53° 47′ 88″ E. of N. 7. 4.

8. 2, 2 ; $x+y=4$, $x-y=0$.

11. $x+a+b=0$, $4a$, $4b$ being the latera recta.

14. $\frac{(a^4y^2+b^4x^2)^{\frac{3}{2}}}{a^4b^4}$. The radius of curvature.

VII. Mechanics.

1. $\frac{1}{2}$ lb., 1·118 lbs., 45°. 4. Angle of inclination of rod $=\tan^{-1}\mu$.

6. Unaltered. 8. The particle projected in direction AT'.

10. Whole work done by the man $=Rh$.

NOVEMBER, 1898.

I. Part I. First Paper.

1. $7\frac{1}{9}$. **2.** £2. 18*s*. $10\frac{1}{2}d$. **3.** $4\frac{1}{4}$ %. **4.** 300 lbs.

5. 11, 24, 39. **6.** $125\frac{7}{12}$. **8.** $x+1$.

9. (i.) $\dfrac{a^6+1}{a(a^8-a^4-2)}$; (ii.) 3, 1; (iii.) $2\frac{1}{3}$, $1\frac{14}{39}$.

10. $\dfrac{(n-1)p-(m-1)q}{n-m}$, $\dfrac{q-p}{n-m}$; first term, 11; common diff., 3; or first term, $-\frac{37}{49}$; common diff., $\frac{3}{49}$.

11. 34. **12.** 5·0003199.

II. Part I. Second Paper.

8. 324 sq. yds. **10.** 972 cub. ft.

III. Part I. Third Paper.

1. $22\frac{1}{2}°$. **2.** $\text{cosec}^{-1}\frac{5}{4}$.

3. $-300°$, $-210°$, $-120°$, $-30°$, $60°$, $150°$, $240°$, and $330°$.

4. $1+2\cos A$. **6.** 120°.

7. ·8987253, $\bar{3}$·0370280, $\bar{1}$·9811057. **8.** 30.

IV. Part II. First Paper.

1. (i.) $x=\pm 2, \pm\dfrac{5}{\sqrt{3}}$; $y=\pm 3, \pm\dfrac{1}{\sqrt{3}}$; (ii.) $x=\pm 3, y=2$; $x=2, y=\pm 3$.

2. $-\dfrac{(n-1)(2n-1)(3n-1)\ldots(\overline{m-2}\,n-1)}{\lfloor m-1}\cdot\left(\dfrac{x}{n}\right)^{m-1}$.

3. $n\pi-\frac{1}{2}(\alpha+\beta)$. **4.** 3·0505245 sq. ft.

6. The locus is a straight line parallel to *BC*.

V. Part II. Second Paper.

1. 16, 20 lbs. **2.** $AD=4·59$ in., $AF=3·915$ in.

4. $7\frac{1}{2}$ inches from the 11 lbs. weight.

5. $5\sqrt{3}=8·66$ lbs., $5\sqrt{7}=13·22$ lbs. **6.** 45°.

7. $180° - \cos^{-1}\frac{u}{v}$ to the direction of motion of the carriage. 8. $\frac{2u}{f}$.

9. $\frac{9}{8}W$ lbs.-wt. vertically upwards; when moving with uniform velocity force $= W$ lbs.-wt.

10. 11 seconds.

11. (a) $2v$ horizontal, $\frac{v^2}{r}$ to the centre; (b) 0, $\frac{v^2}{r}$ to the centre.

12. $a^2 < 4h(b+h)$.

VI. Part III. First Paper.

3. The asymptotes are parallel to AB, AC bisecting the perpendiculars from O on AB, AC.

4. $(a_2b_3 - a_3b_2)(b_1x - a_1y) = a_1(a_2c_3 - a_3c_2) + b_1(b_2c_3 - b_3c_2)$ and two similar equations.

6. $\frac{b(1 \pm e\cos\phi)}{\sqrt{1 - e^2\cos^2\phi}}$. 10. 12·9 seconds.

VII. Part III. Second Paper.

4. $2x - y = 0$. 7. $\frac{a^2 - b^2}{a} \cdot \frac{l(1 - m^2)}{l^2 + m^2}$, $\frac{b^2 - a^2}{b} \cdot \frac{m(1 - l^2)}{l^2 + m^2}$.

8. Tension at $A = \frac{1}{2}$ wt. of board $+ \frac{Wp}{c \sin B}$, where $W =$ wt. of sphere, and p the perpendicular from its point of contact to side BC.

10. $\frac{4m - m'}{4(m + m')} \cdot u$, $\frac{5m}{4(m + m')} \cdot u$.

ANSWERS.

JUNE, 1899.

I. Part I. First Paper.

1. $4\frac{2}{11}$. 2. 8*s*. 5*d*., $2\frac{7}{8}$ %. 3. $208\frac{3}{7}$, 5 ft. 5·03 in.

4. 24 min. 7. $x(x+3)(x-3)(x^2+3x+9)$. 8. $14\frac{2}{3}$, 11 and $-17\frac{1}{3}$.

9. $\frac{1}{3}a-\frac{1}{6}$, $\frac{1}{3}b-\frac{1}{3}$; $100\frac{1}{3}$, $102\frac{2}{3}$.

10. 3430 is 4th term G.P., 3475 is 496th term A.P.; $\pm\frac{a}{2n+1}$.

11. 9, 15.

II. Part I. Second Paper.

1. A rectangle. 3. 2 : 1. 5. $13\frac{3}{7}$ in.

8. 9, 27 sq. inches; 3, 6, 12 and 15 sq. inches. 9. 785 cub. in.

10. $\sqrt{\left\{\frac{8A^3}{27\sqrt{3}}\right\}}$.

III. Part I. Third Paper.

1. 433,380 miles. 2. 2. 3. 70°, 160°, 250°, 340°.

5. Acute $\frac{4}{5}, \frac{3}{5}, \frac{4}{5}$; obtuse $\frac{4}{5}, -\frac{3}{5}, -\frac{4}{5}$. 6. $\frac{1}{\sqrt{2}}, \frac{60}{109}$.

7. 2 : 1, 1 : 3, 3 : 2. 8. *A*, 95° 9′ 37″; *C*, 64° 15′ 23″.

9. 77,613,700,000,000,000.

10. (i.) $ab=1$, (ii.) $ab=x$, (iii.) $a=b^2$, where *a*, *b* are the given numbers.

IV. Part II. First Paper.

1. $acx^2+2(a+c)bx+(a+c)^2=0$. 4. 9 : 2. 6. 86·4307 ft.

V. Part II. Second Paper.

1. *W*. 3. 2·68 inches from *A*. 4. 6 lbs.

6. $\sqrt{2}:1$. 7. $V.\sqrt{3}$. 9. $1\frac{1}{3}$ lbs.

10. 53·76. 11. $\sqrt{\frac{g}{\sqrt{3}}}=4·29$ f.s. 12. $\frac{5}{7}V$.

VI. Part III. First Paper.

3. $\dfrac{aa'+bb'}{\sqrt{(a^2+b^2)(a'^2+b'^2)}}$, $(3\pm 4\sqrt{3})x+(4\mp 3\sqrt{3})y+7\pm\sqrt{3}=0$.

4. 3. **5.** $(1+m^2)(x^2+y^2)-a(x+my)=0$. **12.** $\sqrt{gr}$.

VII. Part III. Second Paper.

2. The angle lies between 0° and 60°. **3.** $6x+7y=11$, $4\frac{1}{3}$.

4. $4x+3y=75$ or $4x+3y+25=0$.

5. cr. $(1-1)$, radius 1; the origin and the pt. $-\frac{8}{25}, \frac{6}{25}$.

6. $y=\pm(x+1)$. **7.** $(a^2-c^2)x^2+a^2y^2=a^2(a^2-c^2)$. **8.** 6.

9. If A, B, C be the corners at which the tensions 1, 2, 3 act respectively, the weight must be placed at the mid. pt. of CD, where $AD=2DB$.

10. $\frac{1}{4}W$. **11.** $\sqrt{\frac{3}{g}}$ seconds. **12.** $\frac{2}{3}\cdot\frac{m_1m_2}{m_1+m_2}\cdot g$.

NOVEMBER, 1899.

I. PART I. FIRST PAPER.

1. $\frac{8}{105}$, 88. 2. £3. 6*s*. 8*d*. 3. £500. 4. 329, 42.
5. 20 miles. 6. 16, 10. 7. $x^2 - 12x + 35$.
8. (i.) 13; (ii.) 3, $-1\frac{1}{3}$. 9. $x = a + b$, $y = a - b$.
10. 325, 9. 11. 225. 12. 70, 4.

II. PART I. SECOND PAPER.

3. 3 : 1. 5. 14 inches. 8. $3CQ = 5QD$; 5 : 16 and 3 : 16.
9. 18 ft., 105 cub. ft. 10. 4 sq. ft. 64 in., 1280 cub. in.

III. PART I. THIRD PAPER.

3. $-\frac{3}{5}, -\frac{4}{5}, -\frac{4}{3}$; $(2n+1)180° - D$. 5. 4 ft. 5·7 in.
7. 20° 36′ 35″, 53° 7′ 48″, 106° 15′ 37″.
8. $13\frac{1}{4}$ miles, 13° 6′ 33″ west of north. 9. 1·500741.
10. (i.) 3·48429, (ii.) − ·283211.

IV. PART II. FIRST PAPER.

1. (i.) $\dfrac{b^2 - 2ac}{a^2}$; (ii.) $\dfrac{1}{a^4}(b^4 - 4ab^2c + 4a^2bd + 2a^2c^2 - 4a^3e)$.

2. $\dfrac{\lfloor n}{\lfloor r \, \lfloor n-r}(3^r - 2^r)x^{n-r}$.

4. $2sm + \dfrac{m^2s^2}{\sqrt{s(s-a)(s-b)(s-c)}}$, $m^2\left\{\dfrac{s^2}{\sqrt{s(s-a)(s-b)(s-c)}} - \pi\right\}$.

8. 102 cub. inches.

V. PART II. SECOND PAPER.

1. $3AC$, acting parallel to AC at D, where $BD = 2AD$. 2. 30°.
3. 8·33, 6·01 lbs.; 33° 40′ to horizontal.
4. 3 lbs. acting downwards, 6 ft. 5. $1\frac{1}{2}$, 5 in. 7. $-\frac{2}{3}$ c.s., $\frac{8}{3}$ c.s.s.
8. At heights 64, 96, 96, 64, 0 ft., greatest height, 100 ft.; 40 ft.
9. (i.) 15 grs.-wt.; (ii.) 60° to the downward drawn vertical.
10. 7 lbs., 32 f.s.s. 12. 3072 f.s., ·007446 f.s.s.

WOOLWICH ENTRANCE EXAMINATION.

VI. PART III. FIRST PAPER.

2. A hyperbola.

3. $x^2+y^2-2ax-\frac{a^2+2c^2}{c}\cdot y+a^2=0.$

4. $2ax-2(b-c)y=a^2+bc-b^2.$

6. A parabola.

8. $2\sqrt{2}$ lbs.-wt.

9. $\frac{2\sqrt{P^2-F^2}-W}{2F}.$

VII. PART III. SECOND PAPER.

4. $\frac{2\sqrt{h^2-ab}}{a+b}, \frac{2h}{a-b}.$

5. $\frac{1}{2}, -\frac{1}{2}; x+y=0.$

6. $b-a, \pm 2\sqrt{ab}.$

8. $\frac{l-2h}{2h}\cdot w.$ If from the end of the ladder a line be drawn making the angle of friction with the vertical and meeting the vertical through the C.G. of the ladder in A, the line joining A to the point of support is the direction of the force.

9. $6:1, e=\frac{3}{4}.$

10. $2w, w$; $\frac{1}{2}r$ above the centre.

11. £825.

12. $\frac{a\{(1+r^2)(1-2r^{3n}+r^{4n})-r(1-r^{4n})\}}{(1-r)(1+r^2)(1-r^{3n})}.$

ANSWERS.

JUNE, 1900.

I. PART I. First Paper.

1. $1\cdot6$, $\cdot01\dot{3}$. 2. £20, £66. 13*s*. 4*d*. 3. £18. 5*s*.

4. 1 ft. 2 in. 5. 12 per cent.

6. $4\frac{1}{2}$ and 6 kilometres per hour. 7. $14x^2-81x+22$.

8. (i.) 7; (ii.) -2, $\frac{1}{5}$. 9. $\frac{c-a}{c+a}$, $-\frac{c-a}{2(c+a)}$.

10. 3069. 11. $\frac{\underline{|20}}{\underline{|15}\,\underline{|5}}=15504$, $\frac{\underline{|20}}{(\underline{|5})^4}$ 12. 16.

II. PART I. Second Paper.

2. The mid-point. 7. Make the angle $ABP=$ the angle C.

8. 89 inches. 9. 1326 sq. inches. 10. 93675 cu. metres.

III. PART I. Third Paper.

1. 15°, 15′, 15″.

2. Sine $\pm\frac{12}{13}$, tangent $\pm\frac{12}{5}$, cotangent $\pm\frac{5}{12}$, cosecant $\pm\frac{13}{12}$, secant $-\frac{13}{5}$.

3. 52 ft. 4. $2n\pi+\frac{\pi}{6}$. 5. $X^2+2XY\cos\phi+Y^2$.

7. (i.) $27\cdot0063$ sq. ft.; (ii.) $5\cdot10626$ ft. 8. 54° 37′ 24″.

9. $\cdot3010300$, $2\cdot3025851$, $\cdot6931471$. 10. (i.) 2; (ii.) 43; (iii.) 42.

IV. PART II. First Paper.

1. $v-u$, $(n-1)v-(n-2)u$, $\frac{n}{2}\{(n-1)v-(n-3)u\}$.

2. $p(p+q)(p+2q)\ldots(p+\overline{r-2}q)x^{r-1}\div\underline{|r-1}$.

6. (i.) $447\cdot87$ links; (ii.) $122\cdot235$ links.

8. $684\cdot233$ sq. inches. 9. 2.

V. PART II. Second Paper.

1. $R\sin\theta$. **2.** 34·8 lbs. **3.** $8\frac{1}{3}$ lbs.

4. $W\sin(45^\circ \pm \alpha)$. **5.** $56\frac{1}{2}$ lbs. **6.** $\frac{\sqrt{61}-2}{1}$ of length of side.

7. $\frac{v^2}{2f}$. **9.** $\sqrt{\left\{2\left(g\sin\alpha - \frac{F}{m}\right)h\operatorname{cosec}\alpha\right\}}$, $\frac{mg\sin\alpha - F}{mg\sin\alpha + F}.h$ vertically.

10. (i.) $\sqrt{\frac{m-m'}{m+m'}\cdot gh}$; (ii.) $\frac{\sqrt{(m^2-m'^2)g.h}}{2m}$.

VI. PART III. First Paper.

3. $\frac{13}{5}$. **4.** $\frac{2}{7}\sqrt{3}, \frac{4}{7}$. **10.** Stable.

11. 30 miles an hour. **12.** $\frac{9\sqrt{3}}{19}$.

VII. PART III. Second Paper.

3. $(\overline{m+nk}-2mn)x+(m+n-2k)y=0$.

4. $x^2+5xy+2y^2-7x-9y+8=0$; $x-y=0$, $7x+9y=8$.

5. $x^2+y^2-2a(x+y)+a^2=0$; $\frac{1}{2}\sin^{-1}\frac{1}{8}$ and $\frac{\pi}{2}-\frac{1}{2}\sin^{-1}\frac{1}{8}$.

7. $b^2x+a^2y\tan\theta=0$. **9.** $\frac{m'v}{m+m'}$, at angle α with line joining mm'.

10. 10 lbs., $1\frac{1}{4}$ lbs.

11. (i.) $\sqrt{gl}$; (ii.) $\frac{1}{2}\sqrt{gl}$ vertical, $\frac{1}{2}\sqrt{5gl}$ at an angle $\tan^{-1}2$ to the vertical.

12. A parabola.

NOVEMBER, 1900.

I. PART I. FIRST PAPER.

1. $29\frac{3}{7}$. 2. ·28845. 3. $12\frac{8}{11}$ acres, £249. 19*s*. $1\frac{1}{11}$*d*.

4. £116. 13*s*. 6*d*. 5. 6·3, 31·5 seconds. 6. 721.

8. 0, $4\frac{3}{5}$; $c=\frac{125}{12}$. 9. $\frac{b+c-a}{a+b-c}$, $\frac{a+c-b}{a+b-c}$.

10. $a^{\frac{2}{3}}-b^{\frac{2}{3}}$.

12. $1-\frac{1}{2}x^2+\frac{3}{8}x^4-\frac{5}{16}x^6+\frac{35}{128}x^8-\ldots$ $\frac{1}{x}-\frac{1}{2x^3}+\frac{3}{8x^5}-\frac{5}{16x^7}+\frac{35}{128x^9}-\ldots$

When $x=10$, value = ·099504.

II. PART I. SECOND PAPER.

2. 100°. 3. 5·3, 1·7 inches. 7. The tangent from *O* to the circle.

8. 156 sq. ft. 9. 2 *r*. 10. 186 lbs.

III. PART I. THIRD PAPER.

1. 206264·8 seconds. 2. $n\pi+(-1)^n\frac{\pi}{6}$. 3. 1; 120°.

4. 45°. 5. $abc\sin\alpha\div 2\sqrt{s(s-a)(s-b)(s-c)}$.

6. $-150°$, -1. 7. 81·4885 ft.

8. $c\sin A\,.\sqrt{a^2-c^2\sin^2 A}$. 9. (i.) 4; (ii.) -4. 10. 1·608.

IV. PART II. FIRST PAPER.

1. $(x+y)(y+z)(z+x)$. 2. Between -3 and -2, 0 and 1, 1 and 2.

6. 2504 ft. 8. A straight line.

V. PART II. SECOND PAPER.

3. 8·6 lbs.-wt. 4. 7 lbs.-wt., 9 lbs.-wt. 5. 4·76 lbs.-wt.

6. 69 lbs.-wt., 30 lbs. 7. 2·07 ft. per sec. 9. $\frac{1}{2}g$, $\frac{1}{8}mg$.

10. $\frac{1}{4}\sqrt{65}$ lbs.-wt. acting at an angle $\tan^{-1}8$ with the direction of motion; 9 : 7.

11. 29·3 ft. before. 12. $\sqrt{\frac{gl\sin^2\alpha}{\cos\alpha}}$.

VI. PART III. FIRST PAPER.

2. A hyperbola with the given points as foci.

3. $\dfrac{p\sin\beta - q\sin\alpha}{\sin(\beta-\alpha)}$, $\dfrac{q\cos\alpha - p\cos\beta}{\sin(\beta-\alpha)}$.

4. $\dfrac{[(p+q)(pr^2+qs^2) - pq\{(\alpha-\gamma)^2+(\beta-\delta)^2\}]^{\frac{1}{2}}}{p+q}$.

5. $90° - \alpha$ with Ox;
$\dfrac{(g\cos\alpha + f\sin\alpha)\{g\sin\alpha\cos\alpha + f(\sin^2\alpha - 2)\} + c\sin\alpha}{2(f\cos\alpha - g\sin\alpha)}$.
$\dfrac{(f\sin\alpha + g\cos\alpha)\{f\sin\alpha\cos\alpha + g(\cos^2\alpha - 2)\} + c\cos\alpha}{2(g\sin\alpha - f\cos\alpha)}$.

7. (i.) $(ax^2+by^2-1)(ap^2+bq^2-1) = (apx+bqy-1)^2$;
(ii.) $4ab(ap^2+bq^2-1) = \{ab(p^2+q^2)-(a+b)^2\}\tan^2\alpha$, and $ap^2+bq^2 > 1$.

8. $\dfrac{Wa^2\sin\theta}{\sqrt{b^2-4a^2\sin^2\dfrac{\theta}{2}}}$. **10.** v, 0. **11.** 60 miles an hour.

12. 39·144 in., $85200\dfrac{h}{r}$.

VII. PART III. SECOND PAPER.

3. The st. line $3x - 11y = 0$.

4. $8x^2 - 2xy - 15y^2 + 11y - 2 = 0$, $8x^2 - 2xy - 15y^2 = 0$.

5. $x - h + (y-k)\tan\alpha = 0$.

9. (i.) $m' > m(\sin\alpha + \mu\cos\alpha)$,
(ii.) $m(\sin\alpha - \mu\cos\alpha) > m$;

acceleration s $\dfrac{m' - m(\sin\alpha + \mu\cos\alpha)}{m+m'}g$, $\dfrac{m(\sin\alpha - \mu\cos\alpha) - m'}{m+m'}g$.

ANSWERS.

JUNE, 1901.

I. ARITHMETIC. (1.)

1. £5830. 3*s*. 2*d*. **2.** 11 cwt. 12 lbs. **3.** 148260.

4. 51 kilom. 840 m. **5.** 2 and 17. **6.** 1800.

7. $\frac{352}{416}$. **8.** $13\frac{47}{72}$. **9.** $2\frac{37}{42}$.

10. $7\frac{1}{2}$. **11.** $\frac{15}{16}$. **12.** 79·55. **13.** 2·485.

14. 1·232. **15.** ·045. **16.** ·6875. **17.** $\frac{13}{2000}$.

18. 1 m. 2 f. 38 po. 3·96 yds. **19.** £3. 0*s*. $9\frac{3}{4}d.$

20. £5. 9*s*. 7*d*. **21.** £27. 11*s*. **22.** £4384. 16*s*.

23. 39. **24.** 3. **25.** 225.

II. ARITHMETIC. (2.)

1. ·508857. **2.** £17. 10*s*. **3.** 1·02 gram.

4. £5. 2*s*. 7*d*. **5.** £65. **6.** Yes.

III. PART I. First Paper.

4. 30°. **9.** A circle, radius 6 cm. or $2\frac{1}{8}$ in. **10.** 2246 cub. yds.

11. 20,000 sq. m. **12.** 108°; ·6881909; $\frac{3}{4}$ per cent.

IV. PART I. Second Paper.

1. 2·72, 7·39. **2.** $x = 2{\cdot}8$, $y = 2$. **3.** ± 2, $\pm\frac{1}{2}$.

4. $2\frac{1}{2}$; 1 and 2, or 2 and 1. **5.** 500.

6. $(2x+y-2)(x+y-2)$. Roots 1, 0; 1, 1; 2, −2; and 4, −2.

8. $1+19x+171x^2+969x^3+3876x^4+\ldots$.

9. sine ·96, cosine ·27. **11.** 17·416 ft. **12.** 54° 8′ 48″.

V. PART II. FIRST PAPER.

1. Quotient x^2-px+p^2-q, remainder $(a+2pq-p^3)x+b-p^2q+q^2$. Conditions $a+2pq-p^3=b-p^2q+q^2=0$.

3. (m^2+n^2) cm.; 25, 312, 313 cm.

5. $n\pi+\frac{3\pi}{4}$.   6. 304·409, ·0199102.

7. $\sqrt{\frac{2}{3}}$ of an edge, any point on the line through the centre of the circum-circle perpendicular to the plane.

VI. PART II. SECOND PAPER.

2. 24·18 lbs.-wt., $\angle AOF=347\frac{1}{2}^\circ$.   4. 9600 lbs.-wt.   6. 299·1 lbs.

7. $AP=1·95$ in., $CP=2·84$ in.; $AQ=1·67$ in., $CQ=3·03$ in.; $AR=1·73$ in., $CR=2·88$ in.

8. 6·06, 14·25, 37·01, 108·06 ft.   10. $1\frac{3}{8}$ poundal.

11. (i.) 46·19 f.s.; (ii.) 7390·08.   12. 3583·29.

15. 1,620,000 ft.-lbs.   16. 20′ 38″.

VII. PART III. FIRST PAPER.

3. $8\frac{1}{2}$, $4\frac{1}{4}$.   5. $\pm\sqrt{c}$, 0; $x^2+y^2-2ky=c$.

6. $35x^2-74xy+35y^2=-12$, $(7x-5y-4)(5x-7y+3)=-12$; $\frac{4}{1}\frac{3}{1}$, $\frac{4}{1}\frac{1}{1}$.

7. $-\frac{ma^2}{c}$, $\frac{b^2}{c}$; 90°.   8. $a\cos\theta=b\sin^3\theta$, $M\operatorname{cosec}\theta$ and $M\cot\theta$ poundals.

9. ·713 lb., 20° to plane.   10. 30,000 lbs., 5 seconds.

11. $2050\frac{1}{3}\frac{1}{2}$ ft.-lbs.   12. $\sqrt{\frac{6b}{g}}$.

VII. PART III. SECOND PAPER.

4. $\tan^{-1}\frac{\sqrt{5}}{2}$.   6. The parabola $y^2=16a(x+2a)$.

7. $\sqrt{\{a^2\sin^2\theta+b^2\cos^2\theta\}}$.   8. $4\frac{1}{2}$ lbs., $4\frac{1}{2}$ ft.-lbs.

NOVEMBER, 1901.

I. ARITHMETIC. (1.)

1. 11 qrs. 3 bus. 3 pks. **2.** £2. 4*s*. $5\frac{1}{2}$*d*. **3.** 1 ton 16 cwt. 8 lbs.

4. 43700500. **5.** 2, 3, 5, 7, 11. **6.** 210.

7. $\frac{7}{30}$. **8.** $7\frac{29}{45}$. **9.** $\frac{71}{78}$. **10.** $21\frac{2}{3}$. **11.** $2\frac{14}{33}$.

12. 9·758. **13.** 2·781. **14.** 9·042. **15.** 1·435. **16.** $\frac{31}{32}$.

17. ·075. **18.** 58 m. 30 cm. **19.** £1031. 1*s*. 3*d*.

20. £11. 11*s*. 4*d*. **21.** 250. **22.** 57 hrs. 12 min.

23. 160 litres. **24.** 9%. **25.** 323 sq. ft.

II. ARITHMETIC. (2.)

1. 75 kilom. 22 metres. **2.** 2·207 lb.

3. £165, £4290, £5376. 5*s*. **4.** ·791. **6.** 168.

III. PART I. FIRST PAPER.

1. $(x^2 + 1·41)(x + 1·19)(x - 1·19)$.

2. (*a*) 180°, (*b*) 360°; $\frac{n-2}{n}$. 180°, 108°.

3. $x = \frac{a^2 + b^2 - c^2}{2a}$, $p = \frac{1}{2a}\{4a^2b^2 - (a^2 + b^2 - c^2)^2\}^{\frac{1}{2}}$,

area $= \frac{1}{4}\{4a^2b^2 - (a^2 + b^2 - c^2)^2\}$ for all triangles.

4. $60\frac{1}{2}$ sq. yds. **6.** 13·85 cm. or 5·196 in.

9. ·259, ·268. **10.** 6 cm., 28 sq. cm.

IV. PART I. SECOND PAPER.

1. 37 days. **2.** $a = 3$, $b = -\frac{4}{3}$, $c = 3$, $d = -\frac{1}{3}$ or -3, $e = -3$ or $-\frac{1}{3}$.

3. The power of x is $pk + ql - rm$. **7.** 10 cm. **8.** 6·75 cm.

9. 3560·47, 2517·28 miles. **10.** 60°, 3·84143, 4·76205 inches.

V. PART II. FIRST PAPER.

1. £1092. 5*s*. $3\frac{3}{4}d$. 2. $2r \sin \frac{x}{2}$. 3. $\frac{1}{8}$, $\frac{7}{8}$.

4. 1, ·8595582, ·7141308, ·5582995, ·3860146, ·1917536, ·0219501, ·0005295, − ·0207130, − ·0777436, ·5159148, ∞. Value of x, 5°·9.

5. 119·51 sq. in. 6. 118 cub. in., 71°. 7. 3·13. 9. Two.

VI. PART II. SECOND PAPER.

1. 6·74, 5·14 lbs.-wt. 4. 30 lbs., $\frac{3}{16}$ inch. 5. 13°.

6. $P=\frac{1}{4}W$, $60\frac{1}{4}$ lbs. 7. 1528 f.s.

8. 16, 48, 80, 112 ft. 9. $4\frac{7}{12}$ tons-wt.

11. K.E. 36, 0, 100 ft.-lbs.; P.E. 64, 0, 100 ft.-lbs. 12. $4m\pi^2n^2r$.

VII. PART III. FIRST PAPER.

2. 2 and $\frac{18}{13}$.

7. Asymptotes $4x-3y+5=0$ and $12x+5y+13=0$. Axes $56x-7y+15=0$ and $x+8y=0$.

10. $\frac{1}{3}g$, $\frac{2}{3}mg$ poundals.

VIII. PART III. SECOND PAPER.

1. $(3, -1)(3, 3)(1, 1)$. Area 4.

4. $\cos^{-1}\dfrac{(a-a')^2+(\beta-\beta')^2-r^2-r'^2}{2rr'}$. $x^2-2\lambda(x-1)+y^2-2y=0$.

5. Vertex $3a$, 0. Latus Rectum a.

6. 0, $\pm\dfrac{1}{\sqrt{b}}$, and $\pm\dfrac{\sqrt{b^2-2ab}}{(a-b)\sqrt{a}}$, $\pm\dfrac{a}{(a-b)\sqrt{b}}$.

7. Distance from first angular point $=\dfrac{\sqrt{26}}{2}$ side of the square.

8. $\frac{1}{2}W\sec\theta$, where θ is inclination of each rope to the vertical.

10. $\dfrac{1}{2}\left\{a+\sin^{-1}\dfrac{V^2\sin^2 a+hg\cos^2 a}{V^2\sin a}\right\}$.

ANSWERS.

JULY, 1902.

I. PART I. First Paper.

1. 25 inches to the mile, $62\frac{1}{3}$ sq. inches. **2.** $\frac{7}{34}$, $\frac{27}{34}$ litre.
4. (i) -22, (ii) $-5·1$. **5.** 29·389265 in., 59·44103 sq. in.
6. 2·71 or 1·29 in. **7.** 105·516, 849·4. **10.** ·9988.

II. PART I. Second Paper.

1. 11 miles per hour. **2.** $\frac{2x(2x^2-5)}{x^4-5x^2+4}$. **3.** 3, -5.
5. $\frac{63}{65}$, $75\frac{3}{4}°$. **6.** 46°. 3′. 16″, 1·0375.
7. 39·32 in. **8.** Yes. **9.** 80·1 ft.

III. PART I. Third Paper.

1. 4·58 in. **2.** a^2+ap+q, 0. **3.** 6·84 cm., 34·2 sq. cm.
6. 43 gallons. **7.** (i) $\frac{1}{4}$ and -2, (ii) ·31 and $-$·81.
8. ·946 acre, 5 chains 18 links. **9.** Not similar, the 3rd side is 2·29.
10. 73°. 23′. 54″, 81°. 56′. 16″.

IV. PART II. First Paper.

1. 6·5, 8·08 in. **2.** £552. **3.** 15 cm., 15 cm.
5. 21·5 chains E., 9·2 chains N.; $66\frac{3}{4}°$ E. of N., 23·4 chains.
6. 714 cu. in. **7.** $\frac{1}{\sqrt{2}}$ of an edge, 7 cm.

V. PART II. Second Paper.

1. 24·9 lbs., making an angle of 18° with *PA*. **2.** 5·47 tons.
3. 80 lbs. **4.** $5\frac{1}{4}$, $1\frac{3}{4}$ tons. **5.** $1\frac{1}{2}$ kilo per sq. cm.
6. 5·07 tons-wt., 133,865 ft.-tons. **7.** 67,760,000, 11,763,889 ft.-lbs.
8. $65\frac{5}{8}$ lbs., $9\frac{3}{7}$ in., $13\frac{1}{3}$ in. **9.** 3·53 secs., 113·14 f.-s.-s.
10. $25\frac{23}{44}$, $34\frac{1}{11}$ miles per hour, $2\frac{1}{3}$ f.-s.-s. **11.** 100 lbs., 250 lbs.
12. 39°.

VI. PART III. First Paper.

1. 4·12. **2.** 31°. **3.** $\dfrac{m_1x_1+m_2x_2}{m_1+m_2}$, $\dfrac{m_1y_1+m_2y_2}{m_1+m_2}$; 2·23, 5·89.

4. 6, 5, and 5, 6. **5.** Eccentricity not greater than 1; $\frac{3}{8}$ and 2 in.

6. x between 0 and 2, y between $2\pm\sqrt{2}$.

7. $x=0$ cuts in imaginary, $x=2$ in real points, and $x=1$ is a tangent.

9. The trail makes an angle of 163° with direction of the train.

10. 160 H.P., $2\frac{4}{7}$ per cent.

VII. PART III. Second Paper.

2. -2, $1\frac{1}{3}$. **3.** $x^2+y^2-4x-2y=0$, 2·24. **4.** $\frac{3}{4}$, $\frac{9}{4}$.

5. 3·24, -4 and -1·24, -4. **6.** A hyperbola.

9. 54·4, 58·2, 77·3, 35·5 grams. **11.** (i) $\dfrac{BP}{AB}W$, $\dfrac{AP}{AB}W$.

NOVEMBER, 1902.

I. PART I. FIRST PAPER.

1. Army 50, Navy 16, Civil Service 13 %. **2.** 3·04 in. **3.** ·29053.
4. 7·8, 6·3, 8·1 mm. **5.** The correct values are 1·25992 and 1·49831.
6. 35 sq. km. **7.** A circle. **8.** ·05 in. error, ·003 in.
9. 21·46 %. **10.** 1·0219865, 5·87354.

II. PART I. SECOND PAPER.

1. 114 days. **2.** 5 : 2 and 2 : 1.
5. 5404 yds. **6.** 104 million gallons.
7. Distances from AB and AC are ·478 and ·533 inches.
9. $£\left(100+\frac{x}{10}\right)$, £350, 4250 copies. **10.** 2·646 in.

III. PART I. THIRD PAPER.

1. £18,524,225. 10*s.* **2.** 3 and 1, if cyclic 2; Yes.
3. ·028, ·0078 kilos. **4.** Latter.
5. 1 + ·15 + ·009 + ·00027 + ·00000405 + ·0000000243 ; £15. 18*s.* 7*d.*
6. 270 cub. ft. **7.** $\frac{ac}{b+c}$, $\frac{ab}{b+c}$. **8.** 3 chains 27 links.
9. 3*s.* **10.** 1620 cub. ft.

IV. PART II. FIRST PAPER.

1. 81,097 miles. **2.** ·00023, – ·00005 ; 279·660302 and ·339698.
4. 2·407 gallons, $a\sqrt[3]{}/2$. **5.** 9·063 sq. in., 3·71 and 5·54 in.
6. $BE^2 : BC^2$. **7.** 11·18 cm., 26° . 34′ and 57° . 33′.
8. (i) The line makes an angle $\tan^{-1} 1{\cdot}2$ with OX, and the readings are 52·8 miles, 3 hours. (ii) 3 hrs. $16\frac{4}{11}$ min., $39\frac{8}{11}$ miles; 2 hours $21\frac{8}{11}$ min. and 4 hrs. $10\frac{10}{11}$ min.

WOOLWICH ENTRANCE EXAMINATION.

V. PART II. SECOND PAPER.

1. $43 \cdot 8$ lbs., $52\frac{1}{3}^\circ$. **2.** 323, 681 lbs.-wt. **3.** 1 ft.
4. 15 lbs. **5.** $\frac{1}{4}W$, $\frac{1}{3}W$. **6.** $\frac{1}{3}$ of an inch.
7. $152 \cdot 7$ ft.-lbs. **8.** 3 lbs.-wt.; $159\frac{3}{8}$, $140\frac{5}{8}$ lbs.-wt.
9. $10 \cdot 8$ ft. **10.** $\cdot 0089$ f.-s.-s., 47382×10^6 lbs.-wt. **11.** $6\frac{1}{4}$ ft.

VI. PART III. FIRST PAPER.

1. $1 \cdot 2$ in. **2.** $8 \cdot 1$.
3. $y = 1 - 2x$ and $\frac{1}{2}(1 - x)$; $\cdot 8$ and $\cdot 45$, $\cdot 6$ and $\cdot 4$, 0 and $\cdot 25$.
6. 11°. **8.** $4 \cdot 39$ f.-s.

VII. PART III. SECOND PAPER.

1. $-1, 2$; $0, -2$; $-\frac{1}{2}$. **2.** $x^2 + y^2 \pm 2by = a^2$. **4.** $x^2 = 2y$.
5. $36x^2 + 100y^2 = 225$; at distances $2\frac{1}{2}$ and $1\frac{1}{2}$ inches from the middle point of AB.
6. 3, 2, 6 ft. from O; 21 ft., 7 f.-s.; $\cdot 004001$ ft., $4 \cdot 001$ f.-s.; 4 f.-s.
7. 79 lbs.-wt. at 45° to the vertical. **8.** 40 f.-s.

ANSWERS.

JUNE, 1903.

I. PART I. First Paper.

1. From Camden, 66·60; from Darlington, 61·40; from Paris 61·36 miles per hour.
2. 6·65 kilometres (or 4·135 miles).
3. 2·6 cm.
4. 63840 cub. ft.
5. 63° 39′ 59″.
6. 365·2422 days, ·0003.
7. $d^2 = 2rh + h^2$.
9. 5·275 ft.
10. 101 ft.

II. PART I. Second Paper.

1. 27 acres.
2. 5′ 45″, 16° 0′.
3. 4·36 metres.
4. (i) $3\frac{1}{3}$, $5\frac{1}{3}$; (ii) 5, 7; (iii) 10, 12.
5. ·6928633.
6. Nitre, 341; charcoal, 68; sulphur, 45 grams.
7. 1·60.
8. a, $-5\frac{1}{8}$; b, $\frac{371}{300}$; £172, £145, £92.
9. 1831·66 sq. ft.
10. 7576 sq. yds.

III. PART I. Third Paper.

1. 196·29 cub. yds.
3. 4207 ft.
4. 16·66 ft.
5. 76·828 chains, S. 72° 38′ 23″ E.
6. £5. 1*s.* $1\frac{3}{4}$*d.*
9. 3951165.
10. ·996, ·985, ·966, ·940, ·906.

IV. PART II. First Paper.

1. 1889·437, 1919·791 cub. inches; 5.
2. 1317·46 yds.
3. 78 chains 44 links.
4. $1 + D^{-1}a + D^{-2}a^2 + D^{-3}a^3 + \dots$, $m\left\{1 + a\left(\frac{1}{D} - \frac{1}{d}\right)\right\}$, $2\frac{1}{4}$ grams.
5. 5·77 ft., 96·225 cub. ft.
7. A right-angled triangle similar to the base, 180°.
8. 645·4 miles an hour.
9. $\frac{12}{13}$.

WOOLWICH ENTRANCE EXAMINATION.

V. PART II. SECOND PAPER.

1. 135 lbs.-wt., making 40° 40′ with *AB*.
2. 7 lbs.-wt. distant $14\frac{2}{7}$ ft. from the larger force.
3. 29·8, 75·4 lbs.-wt.
4. 2 lbs.-wt.
5. 1·8 in. fiom *C* and 1·66 in. from *B*.
6. 1·12 secs.
7. 22·9 lbs.-wt.
8. 1875 ft.-lbs.; 6000 units of momentum.
9. 56·25 lbs.-wt.
10. $1 : 96\pi$.

VI. PART III. FIRST PAPER.

1. $1\frac{1}{3}$, 1.
2. 3·54, − 2·28, and ·34, ·12; length, 4.
5. $x^2 - y^2 = 4$.
6. − ·96, 2·59.
7. 2·3 f.s.
8. 741, 1021, and 696 lbs.-wt.

VII. PART III. SECOND PAPER.

1. $6\frac{1}{2}$.
2. $x^2 + y^2 - 3x - 4y + 4 = 0$.
4. $\frac{5}{4}(x - 3{\cdot}2)$, $x - 3{\cdot}2$.
6. ·16 ft., 2·38 ft.
7. 1° 43′.
8. 24° 14′.

NOVEMBER, 1903.

I. PART I. FIRST PAPER.

1. Lighter.
2. 3.
3. 37·36 and 28·63 ft.; *A*, 132° 34′ 32″; *B*, 20° 55′ 6″; *C*, 26° 30′ 22″.
4. $3\frac{2}{3}$ miles from starting point, 32′ $13\frac{1}{3}$″.
5. 22·9 sq. cm.
7. 2 : 3.
8. $2x^2+(4a-3)x+2a^2-3a+4$.
9. 16 cm.
10. 23661 tons.

II. PART I. SECOND PAPER.

1. £1,612,849. 2*s*. 10*d*.
2. 24 yds., 258·7 sq. yds.
4. 83·97 ft.
5. 9, $\frac{1}{9}$, 243, $\frac{1}{243}$; greater.
6. 4·2 miles an hour.
7. 231 cub. in.
9. 77° 19′ 10″.
10. Volume is 346·7 cub. cm.

III. PART I. THIRD PAPER.

1. £13.
2. 30° and 150°.
3. (i) 8·113; (ii) 2·516.
4. − 16, − 106, 450.
5. 65·286 yds.; error, ·049 yd.
6. (i) ·085 %; (ii) ·18 %.
7. Least number, 60°.
8. 5·94, 4·45; $\frac{3}{4}$.
9. 4·6 cub. ft.
10. 4·2 in.; (i) when *AB* lies on *MN*; (ii) when $\angle MBA = 45°$.

IV. PART II. FIRST PAPER.

1. 29 gallons, 5·15 %.
3. 36·3 %.
4. *A*, *B*, *C*; $\frac{1}{2}x^2-\frac{3}{2}x+1$.
5. 27,894 miles.
6. $x=\frac{2}{3}r$.
7. 8° 35′.
8. 12 ft., 37.

V. PART II. SECOND PAPER.

1. 4·18 lbs.-wt., $50\frac{3}{4}$°.
2. 13·8, 10·8 lbs.-wt.
3. (i) $183\frac{1}{3}$, $116\frac{2}{3}$ lbs.; (ii) $303\frac{1}{3}$, $236\frac{2}{3}$ lbs.
4. 100·14 lbs.-wt.
5. 625 ft., $6\frac{1}{4}$ secs.
6. 2·56 in. from *A*.
7. 3·195 f.s.s., 99·8 lbs.-wt.
8. 500 units of momentum, 5·58 ft.-lbs.
9. 91·3 lbs.
10. ·746.

WOOLWICH ENTRANCE EXAMINATION.

VI. PART III. FIRST PAPER.

1. $3\frac{1}{3}$, $5\frac{2}{3}$, and $5\frac{2}{3}$, $4\frac{1}{3}$.
2. $\frac{44}{13}$, $-\frac{14}{13}$.
5. $2x+4y=1$, $2x-4y+3=0$.
7. (i) 280; (ii) 237·5; (iii) 230 lbs.-wt.; $\alpha°$.
8. 1980 f.s.
9. 29·7, 24·3, 45·9 lbs.-wt.

VII. PART III. SECOND PAPER.

1. $x-y=1$.
2. $x^2-2x-4y+5=0$.
5. The foci are 1·3 in. from centre, and the coordinates of the point of contact are ·706, ·662 in.
6. $757\frac{1}{3}$.
7. $5\frac{9}{11}$, $11\frac{7}{11}$, $2\frac{10}{11}$ f.s.s.
8. 240, 132·7 lbs.-wt.

SD - #0022 - 230724 - C0 - 229/152/20 - PB - 9780282693299 - Gloss Lamination